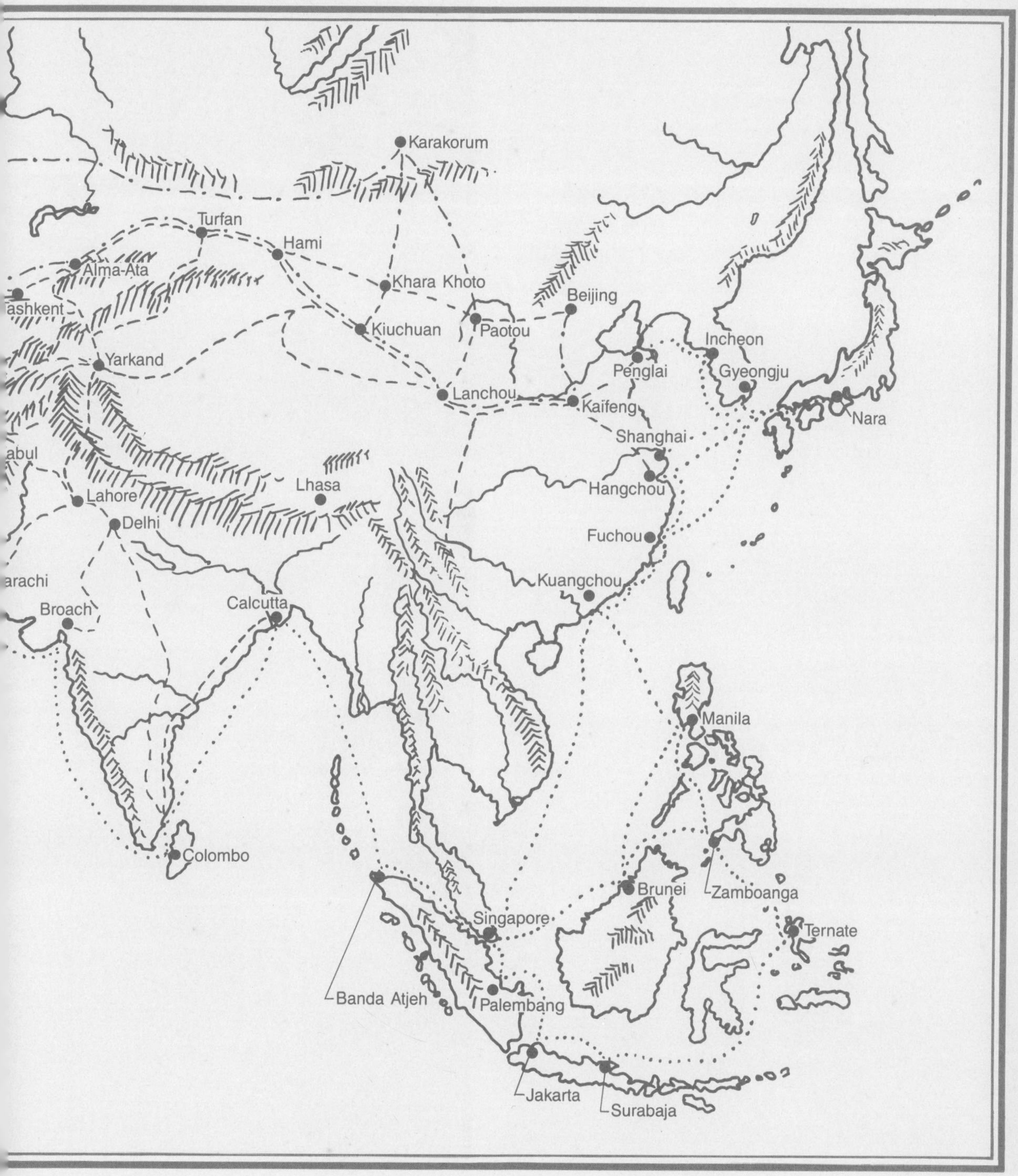

Karakorum
Turfan
Hami
Alma-Ata
ashkent
Khara Khoto
Beijing
Kiuchuan
Paotou
Incheon
Yarkand
Penglai
Gyeongju
Lanchou
Kaifeng
Nara
Shanghai
abul
Lhasa
Hangchou
Lahore
Delhi
Fuchou
arachi
Kuangchou
Broach
Calcutta
Manila
Colombo
Brunei
Zamboanga
Singapore
Ternate
Banda Atjeh
Palembang
Jakarta
Surabaja

THE GENIUS OF JAPANESE CARPENTRY

THE GENIUS OF JAPANESE CARPENTRY

An Account of a Temple's Construction

S. Azby Brown

KODANSHA INTERNATIONAL
Tokyo and New York

In Memory of Maya, Verena, and Uri

Publication of this book was assisted by a grant from the Japan Foundation.

Distributed in the United States by Kodansha International/USA Ltd., 114 Fifth Avenue, New York, New York, 10011. Published by Kodansha International Ltd., 2-2 Otowa 1-chome, Bunkyo-ku, Tokyo 112 and Kodansha International/USA Ltd., 114 Fifth Avenue, New York, New York, 10011.

Printed in Japan.

LCC 88-80126
ISBN 0-87011-897-8 (USA)
ISBN 4-7700-1397-3 (JAPAN)
First edition, 1989

Library of Congress Cataloging-in-Publication Data

Brown, S. Azby.

The genius of Japanese carpentry.
Bibliography: p.
Includes index.
1. Carpentry—Japan. 2. Temples, Buddhist—Design and construction. 3. Yakushiji (Nara-shi, Japan)
I. Title.
TH5608.7.B76 1989 694'.0952 88-27006
ISBN 0-87011-897-8 (U.S.)

CONTENTS

ACKNOWLEDGMENTS

I wish to express my deep gratitude to Tsunekazu Nishioka for his patience, good humor, advice, and support at all stages of this study and to the Reverend Kōin Takada, abbot of Yakushiji temple, without whose permission this publication would have been impossible. Special thanks go also to the Reverend Joukei Fuchigami, formerly of Yakushiji, for providing information concerning the history of the temple, its sect and founder as well as Buddhism in general. I am also grateful to Professor Hirotarō Ōta of the University of Tokyo for his advice and criticism of my manuscript and drawings. I am especially indebted to the the many carpenters, both at Yakushiji and elsewhere, who have patiently answered innumerable questions concerning terminology and procedures and have treated me so warmly.

Sincerest gratitude goes also to the many people who have made my long stay in Japan not only possible but enjoyable, particularly to Professor Hisao Koyama of the University of Tokyo, who generously gave me permission to pursue this study as part of my graduate work, and without whose encouragement and assistance any extended stay in Japan would have been impossible. Thanks are due also to the many organizations upon whose assistance I have relied, particularly the Japanese Ministry of Education, the Japan Foundation, and the Ikeda Construction Company, and to those who made my first trip to Japan possible, especially Atsushi Moriyasu, who shares a deep interest in this subject and who, five years ago, surely could not have foreseen this tangible end-result of a good word placed with the right people at the right time.

Finially, to Lynne E. Riggs and to everyone at Kodansha International, particularly editors Michael Brase and Shigeyoshi Suzuki and designer Shigeo Katakura, whose selflessness and unflagging energy have buoyed me up through the seas of exhaustion.

INTRODUCTION

In the very broadest terms, I have attempted in this book to provide a survey of the unique restoration work and new construction presently in progress at the 1,300-year-old Yakushiji temple in Nara, Japan. More specifically, my aim has been to chronicle the building of one new temple within that compound, the Picture Hall in the Sanzō-in subcompound. The entire project is being carried out under the direction of a remarkable eighty-year-old carpenter, Tsunekazu Nishioka. Part of a line of craftsmen extending back over a thousand years, Nishioka is one the few master temple carpenters, or *miya-daiku*, remaining in Japan. Such *miya-daiku* specialize in the construction and repair of Shintō shrines and Buddhist temples, and in the case of Nishioka his attitude toward his work is deeply influenced by the philosophy of Buddhism. Without him, very likely this book would never have been undertaken, to say nothing of the Yakushiji project itself.

The parameters of the book are shaped primarily by my personal observations of the Yakushiji reconstruction project over a three-year period from 1985 to the present, though my interest in temple architecture dates from the late seventies when I was an undergraduate art student. This is *not* a history book, although history plays an important role; neither is it a "how-to" book, though I would be pleased if designers or carpenters found some inspiration in the descriptions and illustrations contained herein. It is rather an on-the-scenes account of a tremendously complex process of construction, reworked and digested for a lay audience. Since many readers, particularly those with a background in Western crafts or architecture, may be unfamiliar with the design, development, and atmosphere of Japanese temples, I will begin my account with a few general comments.

Temples of Buddhism

A Buddhist temple is a kind of physical model of the abstract philosophy of Buddhism. It is an architectural composition intended to provide a physical and sensual experience conducive to the mental and spiritual experience of the Buddhist faith, with virtually every element designed for both sensual and symbolic value. Admittedly, this relationship is by no means unique to Buddhist temples. Cathedrals, mosques, synagogues, Hindu temples, Stonehenge, Mayan pyramids, and the kivas of the American Southwest, for instance, are at once extraordinary physical experiences and models of a particular cosmology. The uniqueness of Buddhist temples, however, is a function of the way in which the development and spread of Buddhism throughout Asia engendered cumulative architectural change. Thus one can find Buddhist temples that share the same underlying conceptual basis but are nonetheless distinctly Indian, Thai, Chinese, Korean, or Japanese in form and detail.

What then, one may ask, is a Japanese temple? Let me first state that it should be distinguished from a shrine, that is, a Shintō religious structure, although the two may often look quite similar. Second, Japanese temples are almost invariably made of wood, in contrast to Indian and certain Chinese varieties; and they usually, but not always, consist of more than one building, with major structures being considerably larger than the average house. They can be found in the middle of fields and in the mountains; on islands or hemmed in by skyscrapers; pristine or in ruins; gaudy or severe; deserted or thronged with tourists; and, perhaps lastly, contemporary in appearance or more or less ancient in form. And yet, despite this variety, some might agree with a casual acquaintance of mine who remarked, as our train passed through the outskirts of Nara, "You know, the thing about these temples is that they all look alike."

Temples may look alike, but only to the extent that operas or sailboats or plates of pasta look alike. If one is looking for general similarities, they can be found in abundance. If one is alert to differences, there is an unfathomable richness of variety and refinement. It entirely depends on the eye of the observer.

Buddhist architecture came into being as a towerlike tomb called a *stupa*, which was built to enshrine the cremated remains of the historical Buddha, Gautama Shakyamuni, who is said to have lived in the sixth century B.C. It set an important precedent. While the objects of worship were actually the relics hidden deep within, the tower itself became the visible object of devotion, a representation of the Buddha. Its structural configuration and decorations were designed to illustrate certain aspects of Buddhist belief, such as "upward development," "the earth," "the elements," "cycles," and so on. Buddhist sculpture appeared shortly thereafter, evidently influenced by the arrival of Alexander the Great's armies in the Gandara region (now in Pakistan). Craftsmen traveling with the army introduced styles of figurative sculpture, particularly the standing pose, which seems to have been modified slightly and adopted in Buddhist images. The recorded teachings of the Buddha, known as *sutras*, were collected and copied in Sanskrit and Pali, the primary languages of the region, along with monastic rules and commentary. Thus, by about 100 B.C., three primary means of dissemination of the Buddhist faith had arisen: architecture, fine art, and writing.

Buddhism spread to China in A.D. 64 by way of central Asia, with important schools being established in Tibet as a sort of midway point. Chinese civilization had already reached a high level of development and boasted a sophisticated imperial and bureaucratic architecture. These existing forms were adapted for Buddhist use, usually through the inclusion of Buddhist images and decorative details derived from Indian prototypes. The "stacked umbrellas" which crowned the stupa evolved into the pagoda, a wooden or masonry tower with several roofs, and temple layouts were standardized. Distinctively Chinese sculpture and painting appeared about the same time.

The Korean peninsula in the early centuries A.D. was divided into several rival kingdoms, all of which maintained varying degrees of contact with and dependence upon the central Chinese authority. One such kingdom, called Koguryo, accepted Buddhism in A.D. 372, followed by the Paekche kingdom in 384 and Silla in 528. Paekche introduced Buddhism to Japan in 538 when it presented the Japanese ruler with a gilt-bronze statue and a scroll of sutras. The first Korean Buddhist temple carpenter crossed the strait to Japan in A.D. 577, inaugurating the tradition of Japanese Buddhist architecture.

When thinking about the history and culture of Japan, one must always bear in mind its isolation from the Asian mainland and the peculiar cultural relationships that this separation generated, particularly with respect to the Korean peninsula. The strait which divides Japan from the continent proved a formidable barrier to migration and the dissemination of ideas, with the result that the process of civilization was considerably slower than that of the rest of Asia or the Mediterranean basin.

Although scholars disagree, parts of the islands seem still to have been in the neolithic (late stone-age) phase around A.D. 1. While agriculture and metal tools are said to have first appeared in the second and third centuries B.C., iron tools and weapons were not conspicuous until the fourth and fifth centuries A.D. (ironworking itself did not become well established until the sixth century). Unified government, Shintō shrines, and monumental earthwork tombs arose around A.D. 350. The pre-literate phase lasted until the arrival of the scroll and statue from Korea in 538. This evolution, of course, did not occur evenly throughout the archipelago, but by the end of the Asuka period in 661 it is estimated that there were some fifty Buddhist sanctuaries in Japan, scattered from northern Kyushu to the central part of the island of Honshu (the western part of present-day Aichi prefecture).

Continental architecture possessed several features that were radically new. The Japanese had always sunk pillars directly into the ground; Chinese architects set them atop foundation stones and stone-faced podia. The connections between columns, beams, and rafters, which the Japanese had kept as simple as possible—on smaller structures often merely bound with rope—were replaced by intricate bracketing systems. Eaves and other major structural lines, which were unvaryingly straight in early Japanese shrines, now took on the typically "oriental" upswept curve. Roofing tiles made their first appearance, although indigenous thatch and shingle remained in frequent use. And lastly, whereas the Japanese tribesmen had known only bare wood surfaces, the Koreans demonstrated the use of floridly painted wooden members and decorative gilt fittings. We can only imagine how dazzling the new temples must have looked to the average eighth-century villager when he or she entered a new Chinese-style capital for the first time and was confronted with the spectacle of polychrome spires and vast tile-covered roofs extending in an uninterrupted vista to the horizon. The experience must have been awing, inducing an unsettling but intriguing future shock.

The building types introduced by the Korean carpenters at the beginning of the ancient period were of two major

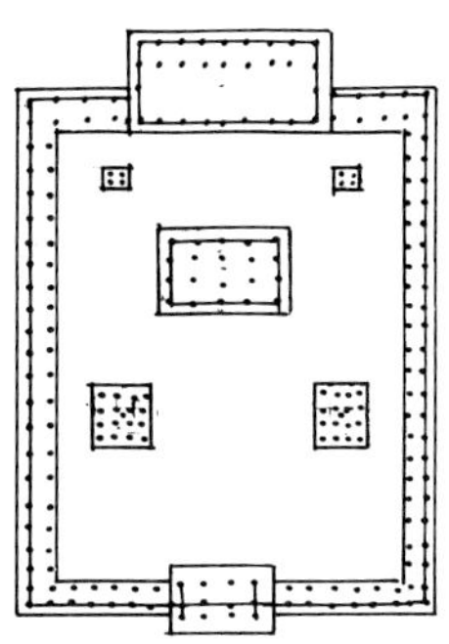

a. Sach'onwang-sa, Silla (Korea), mid-seventh century

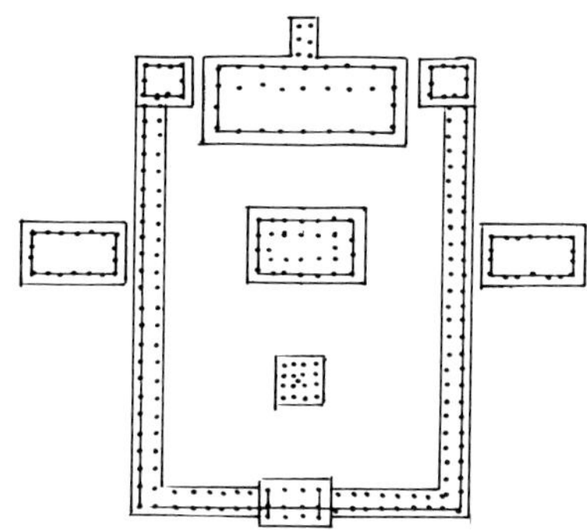

b. Kunsuri Taesa, Paekche (Korea), early sixth century

Scale: 1:2,500

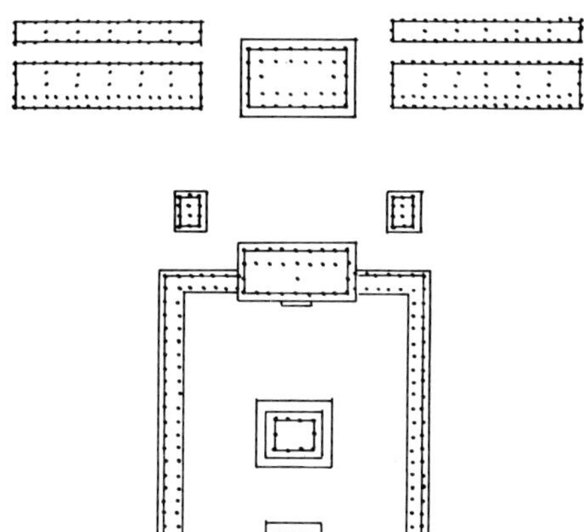

c. Shitennōji (Osaka, Japan), late sixth century

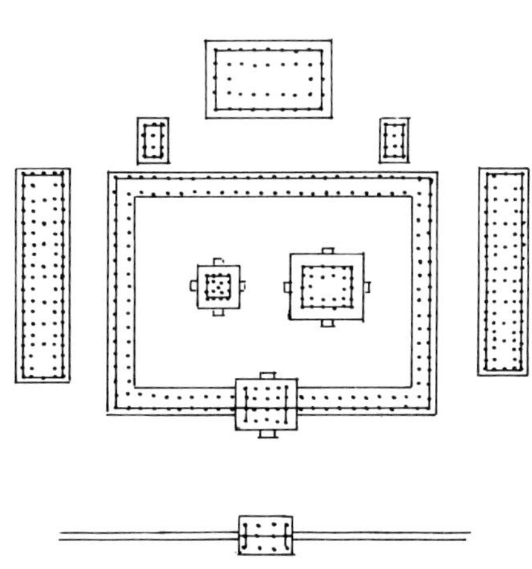

d. Hōryūji (Ikaruga, Nara, Japan), mid-seventh century

1. Early evolution of temple layouts (column placement conjectural for *a* and *b*)

types: the pagoda, or tower (still housing in its base the nominal relics of the Buddha), and the so-called golden hall (housing the most important paintings and statues of the Buddha). These buildings were surrounded by a roofed corridor with a prominent gate, and included several other adjacent buildings (fig. 1). As mentioned above, the earliest

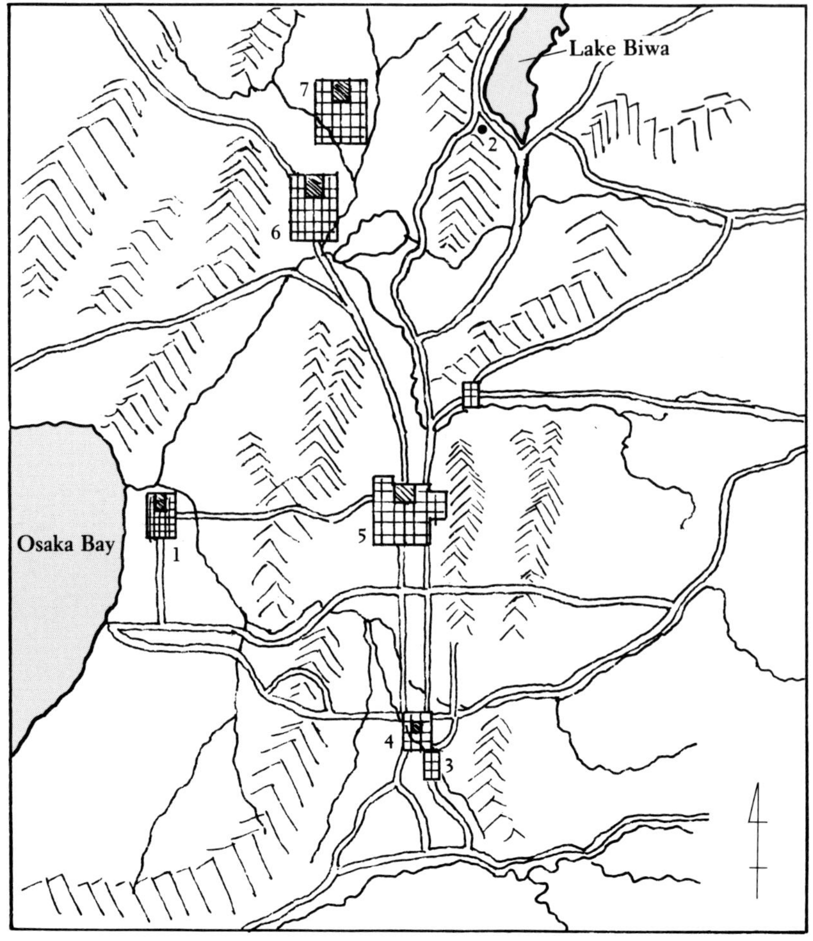

2. Map of the early Japanese capitals: 1) Naniwa, 645–67; 2) Ōtsu, 667–72; 3) Kiyomihara, 672–94; 4) Fujiwara, 694–710; 5) Heijō [Nara], 710–84; 6) Nagaoka, 784–94; 7) Heian [Kyoto], 794–1868.

temples were all built in the area of present-day Nara, in or near one of the capital cities (fig. 2). These earliest temples, which include Asuka-dera and Shitennōji, followed the strict symmetrical layout of their Chinese and Korean prototypes. Although varying in detail, they were all composed to emphasize the soaring pagoda, a means of indicating that the relics of the Buddha in its base were considered more important than his statue housed in the golden hall.

Even at this early time, however, the Japanese seemed to favor asymmetrical compositions, and before long, temples were built to suit this preference. Among the most notable is Hōryūji, established in A.D. 607 by Prince Shōtoku, the figure credited with unifying the Japanese government under one imperial house, partly by encouraging the universal adoption of Buddhism. The present Hōryūji, whose central precinct contains the oldest wooden buildings in the world, dates from A.D. 670, when it was rebuilt after a fire. Despite the widespread approval with which the pleasingly dynamic design of Hōryūji was met, Japan did not completely abandon the symmetrical principle of continental temple layout.

Yakushiji temple, the subject of this book, was erected in 718 (fig. 3). It marked a stunning recurrence of the symmetrical composition of the Chinese prototypes, including two thirty-meter pagodas that flanked and set off the golden hall, an arrangement which may reflect an increase in image worship vis-à-vis reverence for relics. Only the western pagoda contained relics, so the eastern pagoda seems to have been intended primarily as an aesthetic element. Yakushiji was destroyed by fire during the late middle ages, but the eastern pagoda survives to this day intact. In large part, it has served as the basis for the complete restoration and reconstruction of Yakushiji currently in progress.

From these beginnings, Buddhist architecture in Japan proliferated in an evolutionary process that can only be described as confusing at best. New sects, new building sites, shifting patronage, and technological development (and regression) have all contributed to the diversity of building types and details, of scale and experience. And yet, in terms of religious experience, certain factors seem to have remained fairly constant.

3. Yakushiji main compound as seen from the southwest. The reconstructed western pagoda is in the center, the 1,300-year-old eastern pagoda to the right, and the reconstructed golden hall to the left.

Regardless of the specific design, a temple essentially provides a setting for contemplation and prayer, usually in the presence of religious paintings or sculpture. "Good deeds" in the form of offerings of money, incense, and food are performed according to prescribed rituals, and there is usually an area set aside for rest and refreshment. Passage through a temple is a kind of ritual in itself, a process of crossing through successive gates, penetrating deeper and deeper into the temple sanctuary. Temple buildings are almost always set apart on raised platforms which require the ascent of several steps before reaching the doorway. Doors are generally massive with a large sill which must be deliberately stepped over; the size of the doors is partly due to the absence of windows to admit light (although this depends on the specific building). A large sand-filled brazier is usually set before the doorway so that the participant may burn a stick of incense before entering; this is a form of purification. The difference in light levels between inside and outside, one of the most striking features of most temples, is intentional. Due to the broad eaves and relatively small openings for lighting, almost all the light that penetrates the interior is that which is reflected upward from the ground. The effect of this dim light, magnified as it strikes the delicate gilt statuary and fittings, together with the incense-laden air, is deeply moving and somewhat mysterious (fig. 4). When one realizes that this atmosphere

4. The quiet harmony of black lacquer, gilt, and fresh flowers in the dimly lit interior of the Yakushiji golden hall.

has survived unchanged from antiquity, the sense of contact and continuity with the past can be profound. This effect arises from an environment that stimulates all the senses: the coolness of the interior, its dimness and muffled acoustics, the mingled aromas of candle wax and incense, and an almost palpable memory of the tastes of tea, rice cakes, and small temple sweets. That such surroundings are highly evocative of the hope and despair of countless generations and their striving for purity is not incidental. They represent, rather, millennia of refinement and evolution in design, wherein the most profound aspects have been retained and the rest modified slowly over time. This is the consistency that the fellow I encountered on the train in Nara had noted in his remark "They all look the same."

An Unlikely Apprentice

It is not often that a new temple building—to say nothing of a large complex—is built in the "old way." I feel truly fortunate to have had the opportunity to observe such an undertaking close at hand, something which would have been impossible without the permission of master carpenter Nishioka. At the outset, there was no book planned. I approached Nishioka—after a long, labyrinthine process of leads, dead ends, and introductions to people who might obtain introductions to others who might be able to get me an appointment with the master himself—as an awkward American youth with an interest in Japanese carpentry, clutching a handful of slides of my own timber framing work in New England. On learning of my hope to return later to Japan to study more about its wooden architecture, Nishioka offered to take me on as an apprentice. I was shocked, to put it mildly; more shocked, in fact, than flattered, for I knew even then that an apprenticeship, if it was to be worthwhile, should last at least seven years. I doubted I could afford to spend so long in Nara. Besides, absolute obedience is not one of my stronger points.

Despite my refusal of his generous offer, Nishioka helped me receive a grant from the Japanese Ministry of Education for study and research as a graduate student in the architecture department of the University of Tokyo. In addition, he gave me carte blanche to roam around the workshops and construction sites and take photographs. Most significantly, he frequently took time out to answer my questions (which, I fear, served primarily to reveal the true dimensions of my ignorance). I say "answer" my questions, which is perhaps true of the simpler ones about names, dates, terms, and so on, but for the more probing questions—those concerning the "whys" of his motivation and the "hows" of his work—while I usually received something by way of reply, I never got "answers." His responses were almost invariably in the form of subtle hints that I was not being observant enough.

Learning to Observe

In order to know what should be built, Nishioka seemed to say, it was first necessary to observe what already existed. What was worth preserving? What sense, what atmosphere, should be duplicated in a new construction? It is possible for a code or formula to be followed to the letter and yet result in a work devoid of life, inert. This is the meaning of tradition: not formulae, but innate sense; not "design," but patterns of action and use. Only this can result in the preservation of those fragile constructs we call "culture."

In this light, then, what is a temple? Going beyond mere symbolism in a diagrammatic sense, of "trinity" or "planes of existence," a temple must, above all, be a familiar and evocative part of an ancient continuum, a continuum that includes the labor of those who built it. This is what Nishioka and those who share his goals are attempting: to extend that continuum another generation, through their sincere efforts, guided by a sure sense of what has gone before.

Watching this work for the first time is a mystifying and perplexing experience: every action contains an implicit connection with other, unseen actions, a fact which is reflected in the actual configuration of the components. Some have slots, others tenons, and yet others indescribable curves, baffling to the analytical mind. In fact, *every* part, if not itself curved, is part of some greater curve, and fits neatly into, onto, or around one if not several other parts. And yet the carpenter's sole guide, as he negotiates his way through the maze of available lumber, is the information provided him in the form of templates and shop drawings. The fact is, only one person has *all* the information, including that which concerns future stages of the work, and that person is the Master Temple Builder, Tsunekazu Nishioka. The others must concentrate on the job at hand, following instructions. Observation is vital, but wagging one's tongue with "whys" and "whens" is considered not only bad form but a waste of time as well.

When the building was finally erected—within a six-month period in the case of the Yakushiji Picture Hall—most of my questions were graphically and demonstrably answered. The parts literally fell into place. A building, especially a wooden one, is primarily a skeletal webwork illustrating structural stress, generated by desired patterns of human use and occupation. The placement of columns depends upon where people need shelter for standing or moving, and the size of the parts depends upon, say, how large one would like the whole to be. This may be called a "symbolic function," a term that applies to many details of configuration. Then there is the nature of the materials to be taken into account: their properties and limitations, their grain and cellular structure. These and the physical and mental characteristics (including cultural ones) of the builders dictate how the components will be joined to one another. So many patterns. All connected. Like one vast web.

Since so many questions are raised by the enigmatic fabrication of parts, one might expect them to be answered by the actual construction, as in fact they can be. But therein lies a new frustration, for many parts are combined into beautiful new structures only to be obscured by successive additions, and often totally concealed in the end. The wood joints, for instance, are very beautiful in their naked state, but once activated by proper placement, they often become invisible. A line might show on the surface, but it fails to suggest the topological contortions of the whole. The roof structure, equally remarkable in its layers and interconnections, is largely concealed, in deference to an aesthetic code which calls for an illusion of effortless structural support. And yet, it is all very satisfying to note how intricate conformations ultimately fulfill themselves, pieces slipping into allotted slots, craftsmen accomplishing their appointed tasks and moving on.

WOODWORKING IN JAPAN

In most discussions of traditional Japanese architecture, one of the first questions to be raised is why wood has been the primary building material, contrasting so sharply with the Western tradition, whose ancient monuments were almost always of stone and brick, and even with Chinese and Korean architecture, where masonry figures almost as prominently as wood. The Japanese invariably attribute their universal use of wood to the archipelago's natural endowments. Wood and bamboo have existed in abundance from primeval times, while accessible and appropriate building stone was negligible. Seismic disturbances are also frequent. Hence, the reasoning follows, practical limitations dictated an exclusively wooden architecture.

We need not, perhaps, accept this explanation at face value. After all, Japan is a mountainous country with an abundance of rock, and a glance at the stone walks of medieval gardens, or at the meticulous masonry of later castle retaining walls, reveals a highly refined aesthetic sensitivity and technical sophistication. Why were these techniques not utilized in the construction of entire buildings?

It may be that the choice of materials was influenced, much more than is generally recognized, by the manner in which mainland technology was received in Japan. It goes without saying that the sixth-century introduction of Buddhist temple-construction techniques in wood represented a great leap forward in technological complexity, sophistication, versatility, and durability. Building upon woodworking methods whose rudiments were already familiar to the Japanese—we know from archaeological finds that pre-Buddhist Japanese building was chiefly in wood, excepting monumental tombs that often exhibit a precise fitting of huge stones—the Japanese were able to leapfrog several centuries of technological evolution. The important point, however, is that this technology was quickly canonized as emblematic of the new political regime, which sponsored its development and the education of new generations of craftsmen. While there must have been ample opportunity to acquire advanced masonry techniques as well as wood, both during this early stage of Buddhist architecture and in later centuries, the bureaucratic and educational system may have resisted such a step as a costly extravagance or as a challenge to the status quo. If there was thus no clearly perceived need for a parallel stoneworking tradition, the material would be relegated as a matter of course to secondary uses in podia, foundations, decorative lanterns, miniature pagodas, walkways, and, much later, castle walls. Another factor may have been the susceptibility of masonry to earthquake damage, though this would seem to be offset by its increased resistance to fire.

On the other hand, it may have been an element of consideration for the supernatural that influenced these choices. Part of the shadowy value system symbolized by Shintō beliefs stressed love of and respect for wood as a living organism. A form of animism, Shintō ascribes consciousness and personality to natural forces such as wind and rain, sun and moon, geological formations, and nonhuman living things. Certain mountains like Mt. Fuji are held sacred, as are parts of forests and even particular trees. The presence and will of these deities pervaded everything. Without the permission and assistance of the gods, rice would not grow, women would not conceive, and a building would not stand.

These beliefs are celebrated to this day in the form of Shintō ceremonies for planting, marriage, and construction (to name a few); although a diminishing number of people observe the old rituals, natural phenomena are still accorded a measure of respect. The almost religious reverence for wood is, fortunately for us, among the many traditions that have stood the test of time. A tree, like other natural phenomena, is believed to possess a spirit, and a carpenter, when he cuts down a tree, incurs a moral debt. One of the themes that runs throughout Japanese culture is the belief that nature exacts from man a price for coexistence. A carpenter must put a tree to uses that assure its continued existence, preferably as a thing of beauty to be treasured for centuries. There is a prayer that Nishioka recites before laying a saw to a standing tree. It goes in part, "I vow to commit no act that will extinguish the life of this tree." Only by maintaining this pledge does the carpenter repay his debt to nature.

The types of trees that provided the lumber for the ancient temples of Nara and Kyoto continue to preside over the most highly respected traditions of woodworking in Japan. Perhaps the king of them all is the *hinoki*, a species of cypress. It is not only strong and resistant to decay, but easily workable, and has a fine grain, delicate color, and pleasant aroma. Japan's once-extensive *hinoki* forests have largely been cut down, the toll being especially heavy during the mid-century war effort. Today, it is among the most costly of woods. *Keyaki*, or zelkova, is much harder and more elastic, as well as quite durable. Because of its striking grain and color it has traditionally been one of the preferred woods for cabinetmaking, though it also appears in construction.

Sugi, a variety of cedar, is soft, durable, decay resistant, and straight-grained. Employed in both structural and decorative elements, its reddish brown color appears to great advantage in sliding door panels and cabinetry. *Tsuga*, a variety of yellowish brown spruce, is straight, flexible, and dense-grained. It has become the most common building material for all types of wooden construction, which today means primarily houses (although imported Douglas fir may displace it from time to time). *Akamatsu*, sometimes called *mematsu* (female pine), is a red pine. The trunks of this timber, though often twisted, can be used skillfully in roof beams thanks to special techniques that have been developed for layout and joinery. Like most pine, its high resin content acts as a preservative, making it suitable for sills and the like.

Other woods for more specialized uses include *kashi* (Japanese oak). Relatively scarce, it is employed for cabinets and household implements. *Momiji* (one type of Japanese maple) is also used for decoration. *Kiri* (paulownia) is extremely attractive and strong, but unfortunately rare. Although sometimes used in building, it is more commonly found in cabinetry and is the preferred wood for the small boxes in which precious scrolls and pottery are stored. *Kuwa* (mulberry) is hard and dark red, and is found in door-pulls, carvings, and the like. *Kuromatsu* (black pine) and *asunaro* (cypress) are inferior woods of occasional use. *Sakura*, the famed flowering cherry of Japan, is planted for blossoms only; its fruit is inedible and the wood rarely used.

The carpenter is a member of a class of worker known as

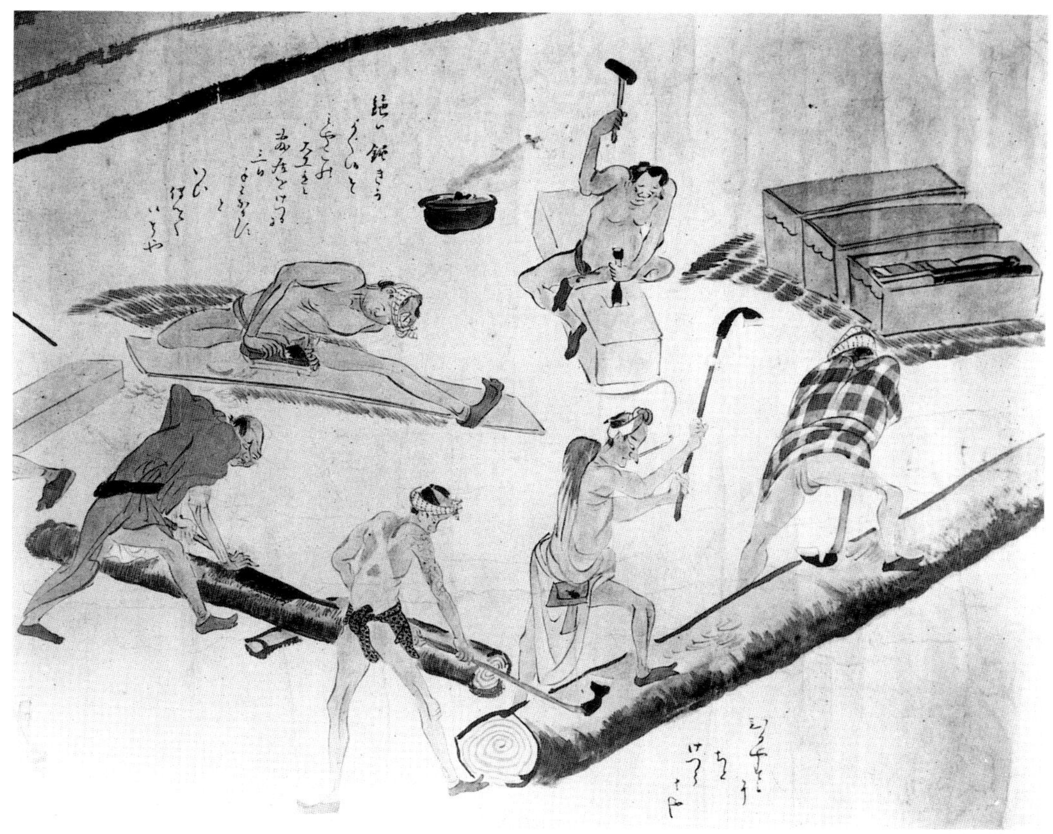

5. Carpenters shown at work in a picture scroll dated 1805. Tokyo National Museum.

shokunin. Usually translated as "craftsman" or "artisan," it has a strong ethical and spiritual nuance in Japanese. The *shokunin*, in addition to repaying a debt to nature incurred by his exploitation of the earth's resources, must fulfill his obligation to society, primarily by doing whatever is required quickly, skillfully, and without waste. This ethical code and social consciousness is cultivated from the beginning of apprenticeship (which before the days of compulsory secondary education started as early as age nine or ten). When a carpenter does a poor job, for instance, it reflects both on the client and on the master; in most cases he is not allowed to redo the work, for that would entail waste. Everyone must live with the result. According to the craftsman's code, this rule is applied to every task in life, even the menial and seemingly insignificant. The carpenter trained according to this tradition will not scoff at the most rudimentary task, says Nishioka, but will do any job that calls for woodworking—even making wooden clogs—with care and precision.

One traditional ideal, then, was the carpenter who could perform any stage of the woodworking process, from felling trees to finishing surfaces, but in actuality there were and are specialities. There are lumberjacks who fell the trees but who are usually concerned only with the main trunk, the direction of fall, and so on. After that the trimmer takes over, removing branches and stripping the bark. Another worker prepares logs for transportation. Finally there is the sawyer, who transforms logs into squared lumber.

Next come the builders, beginning with the *daiku*, or carpenters. Here again, there are several categories. The "house carpenter," highly skilled in comparison to his present-day American or European counterpart, is known for his knowledge of wood and the types of complicated joints still in use. The master house carpenter designs the house itself (according to traditional formulae and rules of thumb) and supervises the fabrication of the components, the erection of the frame, the finishing of the interior, and so on to completion. The actual work of assembling the frame, however, is sometimes carried out by a separate group of craftsmen, the "erectors," who carry out a variety of tasks, including the erection of the scaffolding.

There are also teahouse carpenters, who work on a reduced scale, creating small, finely crafted buildings for the performance of the tea ceremony, usually formulating the design in conjunction with a tea school master. Tea ceremony rules call for subtle, unobtrusive design and an extremely sensitive handling of proportion and detail, as well as studied simplicity. Teahouse carpenters enjoy great prestige.

Temple carpenters such as Nishioka, because of the scale of their work and its sacred nature, perform the most demanding type of carpentry and are regarded with a reverence bordering on awe. Their work includes the design and layout of Buddhist temples and Shintō shrines (although an individual rarely specializes in both). The construction of palaces in ancient times, incidentally, was often on a par with temple building, but the carpenters were segregated into separate guilds. Temple carpenters must be well versed in Buddhist theology and ritual, and are treated as members of the clergy on certain ceremonial occasions. Nishioka estimates that there are some fifty master temple carpenters working in Japan today, though very few have the opportunity to work on major projects.

THE MASTER CARPENTER

Tsunekazu Nishioka is a gentle man. With an eventful eight decades of life behind him, he sits arguably at the apex of the world of Japanese carpentry. He has overseen a major reconstruction at Hōryūji, Japan's oldest Buddhist temple. He has accomplished the better half of the reconstruction of Yakushiji, which is only slightly younger than its illustrious predecessor at 1,300 years of age. In the process, Nishioka has been showered with virtually every major cultural prize bestowed in Japan, both public and private.

Part of his achievement is due to fate, for the position of master carpenter of a major temple or shrine is virtually hereditary, and in 1908 Nishioka was born into the family that had produced master carpenters for Hōryūji temple for five or six generations. Those bearing the Nishioka name represent only one branch of a larger family which claims as an ancestor the builder of Osaka Castle in the early seventeenth century. Hideyoshi, the lord for whom the castle was built, ordered the builder killed because of his knowledge of the castle's secrets. Katsumoto Katagiri, a high-ranking warrior, helped him escape, whereupon he found a quieter life for himself as carpenter at Hōryūji.

Nishioka's father was actually a farmer who had been adopted into a carpentry family as a precaution against starvation at the end of the nineteenth century, which were lean years for Hōryūji. But Nishioka's grandfather was the temple's master carpenter, and he permitted Nishioka to frequent the workshop from the age of three or four. After entering elementary school in 1915, Nishioka helped out with the sweeping, the fetching of tea, and such during his summer vacation. His grandfather insisted, against Nishioka's expectations and his father's wishes, that the boy be further educated at an agricultural school, which he entered in 1921. Grandfather Nishioka reasoned that the

6. Tsunekazu Nishioka at his desk at the Yakushiji work site. Hanging on the wall are numerous wooden drawing templates for plotting roof curves etc.

7. Nishioka inspects his tools.

family should have at least one person capable of farming in times of shortage, and that, in addition, a carpenter should learn as much as possible about wood. As a result, Nishioka devoted four years to absorbing the latest scientific knowledge about earth, fertilizer, forestry, and husbandry.

After finishing school in 1924, the young man expected to begin carpentry immediately, but was told to first spend one year growing rice. Only then, under his grandfather's supervision, was he allowed to handle tools, beginning with learning how to sharpen them. At the same time he was taught how to conduct relationships with other carpenters, and at night he listened to explanations of the carpenters' *kokoroe*, or rules of thumb. These oral teachings, as transmitted by his grandfather, still form the core of Nishioka's values and ethic.

Nishioka worked as an apprentice for about a decade, with an interruption for military service in 1929. Early projects included minor repairs on various buildings at Hōryūji, the complete dismantling and restoration of other temples in the Nara area, construction of a new Shintō shrine, and the erection of a gate to a new palace for a royal prince. In 1931, he made a detailed 1/10-scale study model of the Hōryūji five-story pagoda (now in the collection of the Tokyo National Museum). On some smaller projects, he acted as master carpenter; at other times, he was assistant master. In 1934 he assumed the responsibilities of master carpenter for major dismantling and reconstruction work at Hōryūji, a maintenance operation that is usually carried out every 200 years. This project, called the Hōryūji Shōwa-Era Great Reconstruction, was Nishioka's major formative experience as a master temple carpenter.

During the reconstruction of Hōryūji, Nishioka discovered that the original builders had cut all of their *hinoki* from the hills in the Yamato area (the region surrounding Nara), and that their use of lumber—the matching of a tree's natural qualities with its ultimate placement in a building—were in keeping with the oral precepts that had been handed down to him. "Different mountains make different wood," he says, adding that "building is a matter of matching the individual personalities [of trees as well as carpenters]." Traditional wisdom provides innumerable guidelines for exploiting the unique characteristics of different trees, utilizing twists so that they counteract each other, making sure that trees which grew on the southern face of the mountain are used in the southern side of the building, and so on. "Coming from a family of craftsmen, I had learned these principles already," he says, "but it was only when I took the buildings to pieces that I discovered that all of Hōryūji was constructed in this way. I was extremely moved. The oral tradition had been applied without exception." But the ancient standards were not always maintained, he asserts, and the results can be seen in repair work carried out some six centuries ago. "They simply matched good-looking trees with good-looking trees. When I rebuilt, I reverted to the old ways, perceiving the nature of individual trees, matching them north and south" ("After 51 Years, a Temple Is Restored," *Washington Post*, 26 December 1985).

Thus, fortune was smiling when it fell to Nishioka to oversee this rare dismantling and repair work at Hōryūji, to check up on, so to speak, the work of his predecessors, albeit of a millennium past. "The old builders were people of art who approached their work with religious devotion," he says. "They had no way to know how their creation would fare over the span of 13 centuries. The privilege of finding out fell to me. I felt like a highest-level doctor of anatomy" (*Washington Post*). In his consummately informed opinion, the buildings at Hōryūji are good for at least another 2,000 years, provided they receive regular checkups and proper maintenance.

It was during the early stages of the Hōryūji restoration that he fully realized the benefit of his agricultural education. For one, he was called upon to examine the soil under the foundations of the temple's eastern precinct, an analysis which won him the respect and recognition of his colleagues. He has often maintained that his education has served him well: "These are the things a person needs to know in order to make a house. . . . If he is going to make a wall, he needs to know what kind of earth is good. If he is going to make tiles for a roof, the same holds true. . . . And, of course, if he uses wood, he should know how and why it grows like it does" (Teiji Itō, ed., *Dōkyū no Shokunin* [Tokyo: Heibonsha, 1985], 57).

"The Hōryūji five-story pagoda and the Yakushiji eastern pagoda are both constructed of wood which had been growing for 1,000 to 1,300 years. . . . That means, altogether, each of those trees has been standing for 2,600 years!" He gives as an example: "*Hinoki* is strongest 200–300 years after cutting. . . . This is traditional wisdom, so I was extremely pleased that recent research by Professor Jirō

Kohara [a forestry specialist] backed it up" (*Dōkyū no Shokunin,* 66). A sparkle comes into Nishioka's eyes when he talks about *hinoki* (cypress), his favorite wood. "There are six species of *hinoki* in the world, but Japan's is the best quality. That is why *shiraki zukuri* [bare-wood construction] began here. The six types are Japanese *hinoki*—which was called *ma-ki*, or 'true wood,' by the ancients—*sawara*, an inferior species found in Japan, *tai-hi* [Taiwanese cypress], and three species of North American cypress. . . . The cell structure of Japanese *hinoki*," of which the very best comes from the Yoshino region where it is favored by particularly fine soil, he says, "is very tight. It has a straight grain, nice color, and a delicate aroma. It is stable and easily planed, but it is unforgiving, so one must work it with only the best tools. If a sculptor tries *hinoki* just once, he will never want to touch any other wood" (Jirō Kohara, *Nihonjin to Ki no Bunka* [Tokyo: Asahi Shinbunsha, 1984], 100–101).

Temple construction requires high-grade logs of large girth, which means of great age—ideally the 1,000 or more years that went into the columns of Hōryūji and Yakushiji. But such trees have become extremely scarce in periodically deforested Japan. As a result, Nishioka has made several trips to Taiwan to select wood for the reconstruction of Yakushiji begun in 1970. One of the carpenters' sayings, he relates, is: "Don't buy trees; buy a mountain." Thus the temple has purchased large forested tracts in Taiwan for use in the restoration project. The Taiwanese *hinoki* is very similar to the Japanese variety, but somewhat less elastic, a feature which at times requires structural members—particularly roof beams, which fortunately are hidden—to be made larger in section than traditional design dictates. But age is not always an assurance of quality, he observes. "I saw some 2,000-year-old *hinoki* growing in Taiwan, but their leaves were lighter than other trees only 200 years old. Dark leaves are a better sign. The branches of the trees with the paler leaves were sagging because they were rotten inside" (Tsunekazu Nishioka, *Ki ni Manabe* [Tokyo: Shōgakkan, 1988], 200).

Although Nishioka values the contribution his formal education has made to his development, he regards it as something which, while helpful, is not essential, especially when it comes to training one's sense perception. "Academics," he says, "can be very foolish. They take simple things and make them difficult. . . . Japanese society today measures people by their educational credentials, with the lamentable result that other equally valid ways of learning are being forgotten, even though they're backed by 1,300 years of experimental observation in Japan alone. And this has happened so quickly: in a mere 100 years, thirteen centuries of accumulated knowledge has been allowed to leak out of our culture, not just in architecture, but everywhere."

Because the craftsman's knowledge is based on experience, on endless repetition and refinement, on direct imitation of a master's methods, a technique or attitude disappears with its possessor if not passed on through use. After the demise of a living tradition, one can analyze its artifacts for clues to their creation or try to reconstruct methodology based on writings or hearsay, but something essential is inevitably lost. During the premodern era, there was an active dialogue among Japanese carpenters and a sense of competition which resulted in constant innovation and perceptible improvement in skills for the craft as a whole, all done while struggling against a progressively diminishing quality of available wood. But at that time, wood construction represented the only technology available for building in Japan; nowadays it is used extensively only in prefabricated houses, and in a manner which is increasingly automated. In the face of interlocking industries which have settled upon steel and concrete as the most practical and profitable means of construction, wood has become something of a luxury. Likewise, the traditional means of training carpenters, which had long produced a steady stream of highly skilled craftsmen through a self-regulating apprenticeship system, has now been almost completely dismantled in favor of universal standardized education geared toward the production of office workers and technicians. Carpentry, even temple carpentry, is far from being a lucrative occupation, partly because the Japanese value system has come to disdain manual labor. "Young people? No, they don't want to do this kind of work anymore," observes Nishioka wryly. Most carpenters seem to agree that actual development, in the sense of positive innovation and improvement, is no longer possible in the field. "It is impossible to improve on the ancients," Nishioka insists. "They were incomparably better. The most I can do is to keep a small fraction of their knowledge alive."

Some may be able to accept this decline, this cultural deskilling, without feeling any sense of loss; many Japanese regard it as regrettable but inevitable. Others, however, feel that it is neither inevitable nor inconsequential, but rather a case of painful and unwarranted negligence. The carpenters themselves, including the younger apprentices, carry themselves with the esprit that marks the last of a great breed. To share their life is to share their pain, which they generally keep well hidden except when drunk. One younger carpenter, near the end of a long day of intoxicated celebration, turned to me with a grave countenance and pleaded, "You said it's painful to see how few of us are left. Suffer more, please. Then you might learn how we feel."

Needless to say, crafts are inseparable from the culture in which they are found—they are products of a complex set of interactions, a mirror of social relations. The vigor of any one part is a sign of the vigor of the entire organism, like the color of the leaves on a tree. For instance, Japanese carpenters are dependent upon the finely wrought tools that evolved concurrently with their craft. What happens when the makers of these tools begin to disappear as demand for their work declines? This has been occurring for decades. Carpenters cannot work without adequate supplies of wood, but such supplies are now dwindling throughout the world. Other crafts vital to temple building are suffering similar fates, among them traditional sculpture, painting, and metalworking. During their premodern heyday these crafts formed a coherent whole, a dynamic and interdependent living system not unlike a temple itself.

The master carpenter is the crucial peg in this system. His primary task is to bring together the necessary people and act as the center. He must have extremely broad knowledge, embracing all the related crafts, and a comprehensive viewpoint that holds everything together. If this central peg is removed, the entire system collapses.

In order to maintain continuity with the past, one of the master carpenter's most vital functions is that of education: training those who work under him. In real terms, this means providing them with the best possible example and allowing them to learn through observation and experience. Rarely is anything *explained* fully to the apprentice; he must draw his own conclusions and develop his own instincts. The reasons behind given instructions become evident in time, and no amount of prior speculation or analysis is as effective a learning tool as witnessing an actual process in an alert frame of mind. Some factors, such as the natural settling and movement of wooden structural members over time, can be grasped only after several years of observation, and then only if the carpenter has his original actions continually in mind. Nishioka, for his part, lives in close contact with the entire body of his work, which spans six decades. He is continually forced to assess his successes and failures, seeing where the wood has taken unforeseen twists, where he was hasty or distracted while working.

Consequently, Nishioka is a strict teacher, seeking to form carpenters from the heart outward, as it were. Priority is given to instilling in the young apprentice a sense of respect and humility, not merely toward his superiors, but more importantly toward the wood and the work. The first tasks are deceptively simple: sweeping the shop floor, fetching tea, helping to lift heavy members. Yet even these seemingly mindless tasks have the potential to teach concentration, exactitude, and teamwork, all of which are absolutely essential for the more complicated work ahead. Once an apprentice has begun to develop the proper attitude, he is gradually introduced to tools and begins to assist the senior carpenters with their individual responsibilities. In about seven years, he will be capable of shaping a complex piece from shop drawings and templates; after fifteen years or so, he will know how to make the templates themselves from the master's drawings.

A major project, such as that at Yakushiji, will involve, during its busiest periods, about two dozen carpenters of widely varying age and experience. The core of this crew will be composed of seven or eight workers who have been associated with Nishioka as apprentices for a number of years; other experienced carpenters will be taken on for a year or so. At times, even senior carpenters may leave to work on other projects for a few months, or even years, returning to Nishioka's workshop when they are needed. The result is a loose network of carpenters, sometimes far-flung (but generally within the Kansai area), who form a pool of well-trained labor. And all have been molded in some way by the philosophy that Nishioka expresses so lucidly. While most of the carpenters, and the senior assistants in particular, seem wary of verbalizing their intentions, occasionally philosophical observations do emerge. One day, a thirty-year-old assistant master, seemingly frustrated by my insistent questions about aspects of design related to wood movement, blurted out, "Only three things are important: sun, wind, and rain. If you understand these things, you can understand wood. If you understand wood, you can be a carpenter." On another occasion, a much older carpenter, while walking to lunch, for no apparent reason pointed to a puddle left by the previous day's rains and said, "It's moving. It's *all* moving," a statement rendered even pithier by its brevity in Japanese.

Nishioka now spends almost all of his time on design work, leaving the day-to-day training of apprentices to the senior carpenters. Nonetheless, he periodically checks on everyone's progress, offering his criticism and encouragement. When actually supervising some aspect of the work, he has the bearing of a general in the field, barking orders and accepting no questions. Because of his well-earned status and uncanny perception, even the most senior apprentices are awed in his presence, some of them trembling at the thought of the reprimand he would deal should some barely perceptible flaw be discovered in their work.

On one occasion, a relatively new apprentice related a not-so-traumatic but memorable criticism he had received. "I was planing a block of wood, and I hadn't quite gotten the hang of it. For some reason it wasn't going smoothly. Nishioka was walking across the shop floor about twenty feet away, too far away to see what I was doing, but he called out, 'You're bearing down too hard on the back of the plane. I can tell by the sound.' Sure enough, I adjusted my grip, and the problem was solved." Another carpenter tells how he was taught to distinguish the region in which a particular piece of Japanese *hinoki* was grown on the basis of its smell, with that of northern species being palpably sharper. On another occasion, I watched four senior carpenters standing at attention, silently accepting a stiff rebuke from the master. Their crime: someone had miscalculated a few millimeters on a hip rafter. The difference was hardly noticeable, even close up, but since the beam was designed to achieve its perfect form only after several years of sagging and shrinking, this small error would be magnified and possibly distort the whole. Fumed Nishioka, "They'll laugh at me. They'll say, 'That's not the way a hip rafter should look!' And I won't be around to defend myself."

Indeed, the most remarkable thing about Nishioka is his sense of time, time as measured in centuries and millennia (fig. 8). His viewpoint is purely Buddhist: time is cyclical, the universe is constantly being destroyed and remade, over countless eons, throughout all of which our spirits are intact. Life is painful, especially when compared with the intervals between death and rebirth. "If a carpenter dedicates

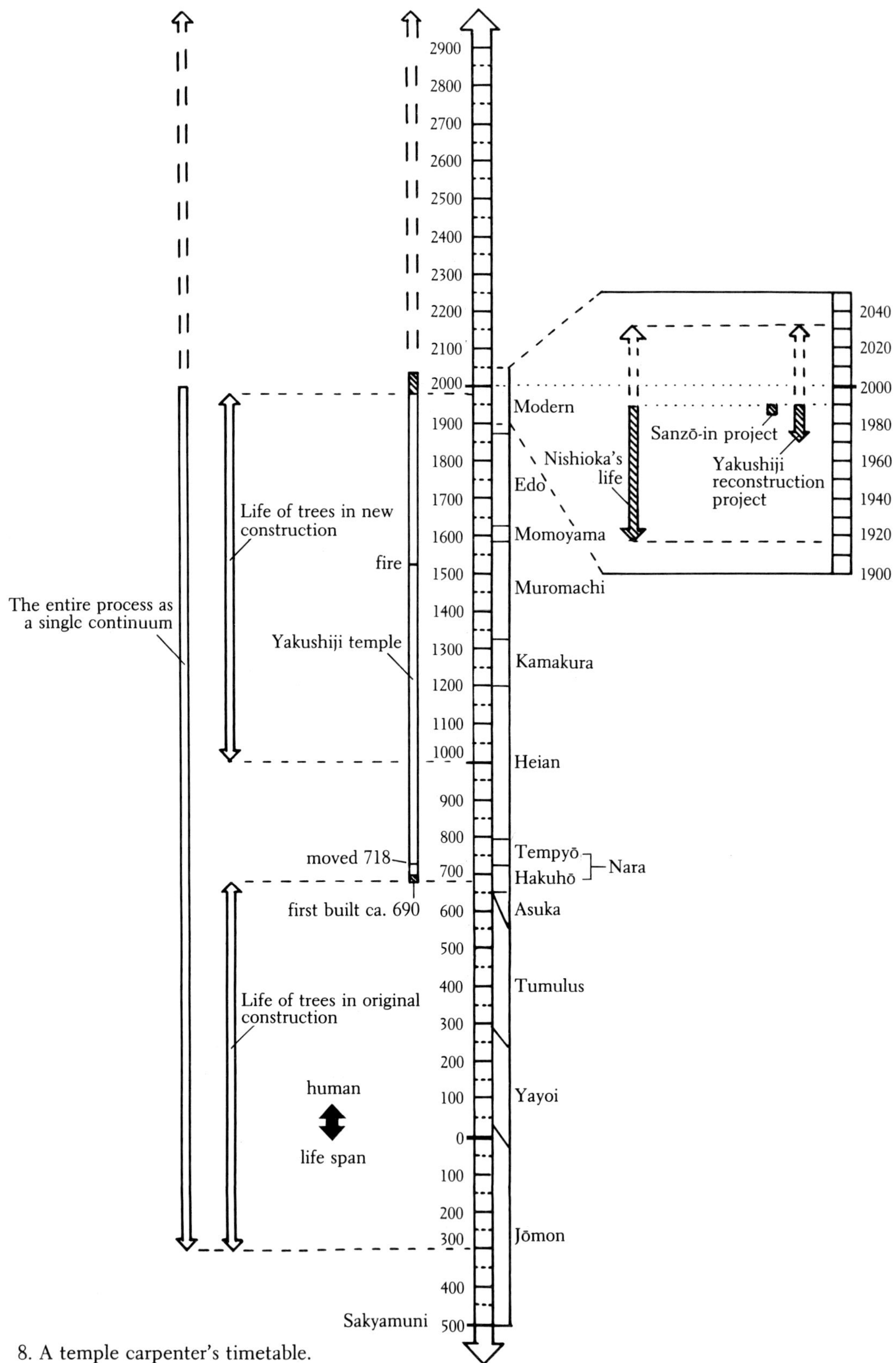

8. A temple carpenter's timetable.

himself spiritually to the construction of a temple, and that temple lasts a thousand years, then he will have a thousand-year interval between this life and the next. Consequently, that thousand years really isn't very long. And when you think that a tree takes a thousand years to grow large enough to use for a temple column, and that the temple may stand for a millennium or more, even a decade spent in its construction is infinitesimal." So, he believes, one should take as much time as necessary. One must concentrate completely all the while, and be in good spirits, because a craftsman's frame of mind permeates every aspect of his work, and the work will bear the imprint of its creator as long as it exists.

Admittedly, it is difficult for most of us to conceive of a thousand years as a brief period, or to believe we can delay our rebirth by dint of diligent work. Nishioka, however, is convinced that the ancient Japanese temples, several of which are over a thousand years old, were built in the spirit he describes, and that the same spirit can be maintained today. This is what he has been attempting at Yakushiji.

THE YAKUSHIJI PROJECT

Yakushiji temple, one of the original seven major temples of Nara, was first planned by the emperor Temmu in A.D. 680, at which time the capital of Japan was a city south of present-day Nara called Fujiwara (now known as Asuka). Temmu's intention was to have the temple dedicated to Yakushi Nyorai, the Buddha of healing, as a means of aiding the recovery of the empress Jitō from a serious illness. As fate would have it, the empress survived, but Temmu himself died before the temple was completed. Empress Jitō completed the construction as well as the casting of the primary bronze statuary. Yakushiji was dedicated in 697 and completed in 698.

A little over a decade later, a new capital city, called Heijō, was built slightly to the north of Fujiwara. This is the site of present-day Nara. Since the jointed wood construction of the day allowed for easy dismantling and transportation of buildings, many major structures, such as palace buildings and temples, were simply moved to the new capital. Yakushiji was relocated to its present site in 718, where it was afforded a large, prominent plot of land, appropriate to its status as a "first-class" temple. Following Chinese precedent, the capital was carefully zoned into blocks, which were awarded to temples, aristocrats, merchants, and others on the basis of both practicality and prestige (figs. 9–10).

Yakushiji has from the outset been a temple of the Hossō sect, which was founded in China in the seventh century by Hsüan-tsang (Genjō Sanzō in Japanese) and was the first Buddhist sect to be brought to Japan. The original layout of the temple, which is a complex of several major buildings grouped within and around a large walled enclosure, features twin pagodas, each thirty-four meters high. Though in fact three stories, the pagodas are designed to appear six-storied through the addition of "false" roofs be-

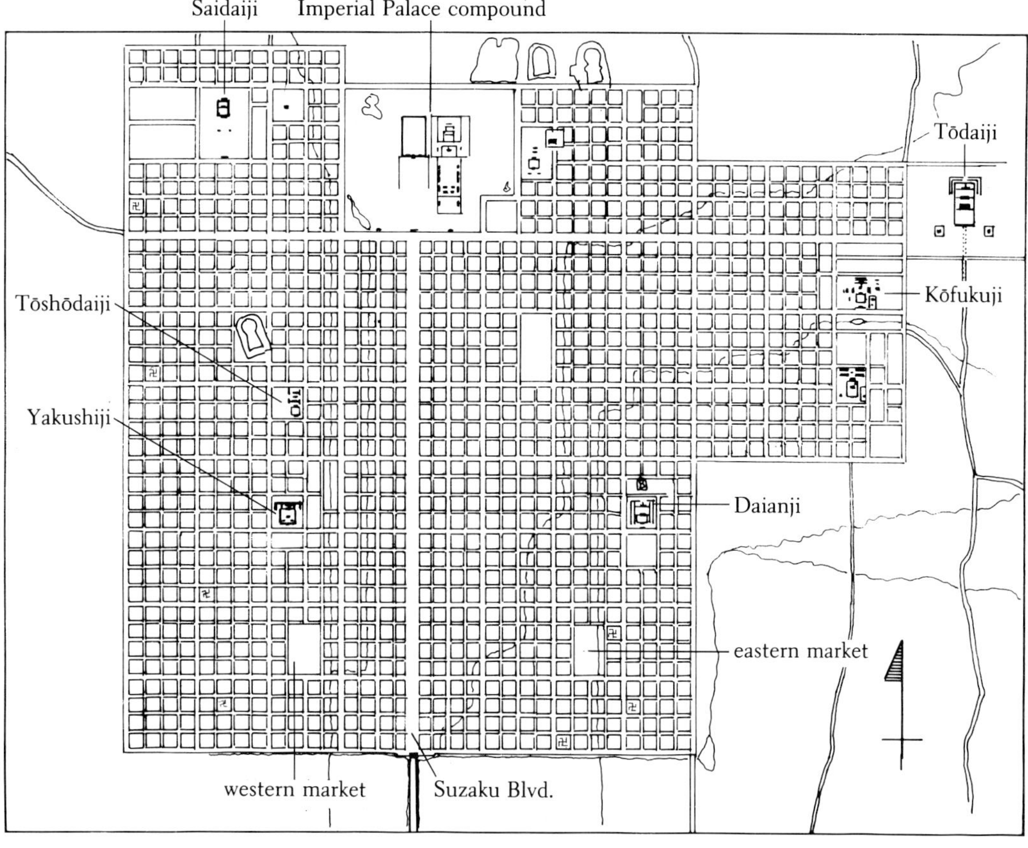

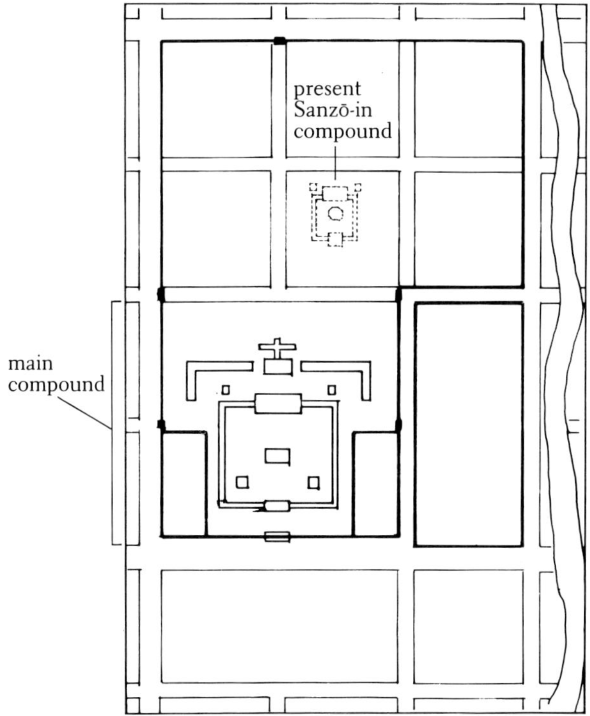

9–10. Map of Heijō (Nara) in the eighth century and detail of Yakushiji area.

11. Eastern pagoda of Yakushiji main compound has survived since A.D. 730.

tween the roofs of the actual structural floors (fig. 11). This design treatment is used in the nearby golden hall and other buildings as well. The golden hall (fig. 12) houses the triad of bronze statues cast late in the seventh century, representing Yakushi Nyorai and two attendants, Nikkō Bosatsu (the Sunlight Bodhisattva) and Gakkō Bosatsu (the Moonlight Bodhisattva), all three of which are now designated National Treasures. The design of the main

12. The golden hall of the main compound as reconstructed (completed 1975).

enclosure of Yakushiji centered on the main figure of Yakushi Nyorai, and the other structures were arranged in such a way as to emphasize its centrality (fig. 13). In its most complete form, the main complex included the two pagodas, the golden hall, a lecture hall, a massive southern gate, a smaller central gate, and a surrounding ambulatory corridor. Among lesser edifices were a bell tower, sutra repository, and residences, offices, and a refectory.

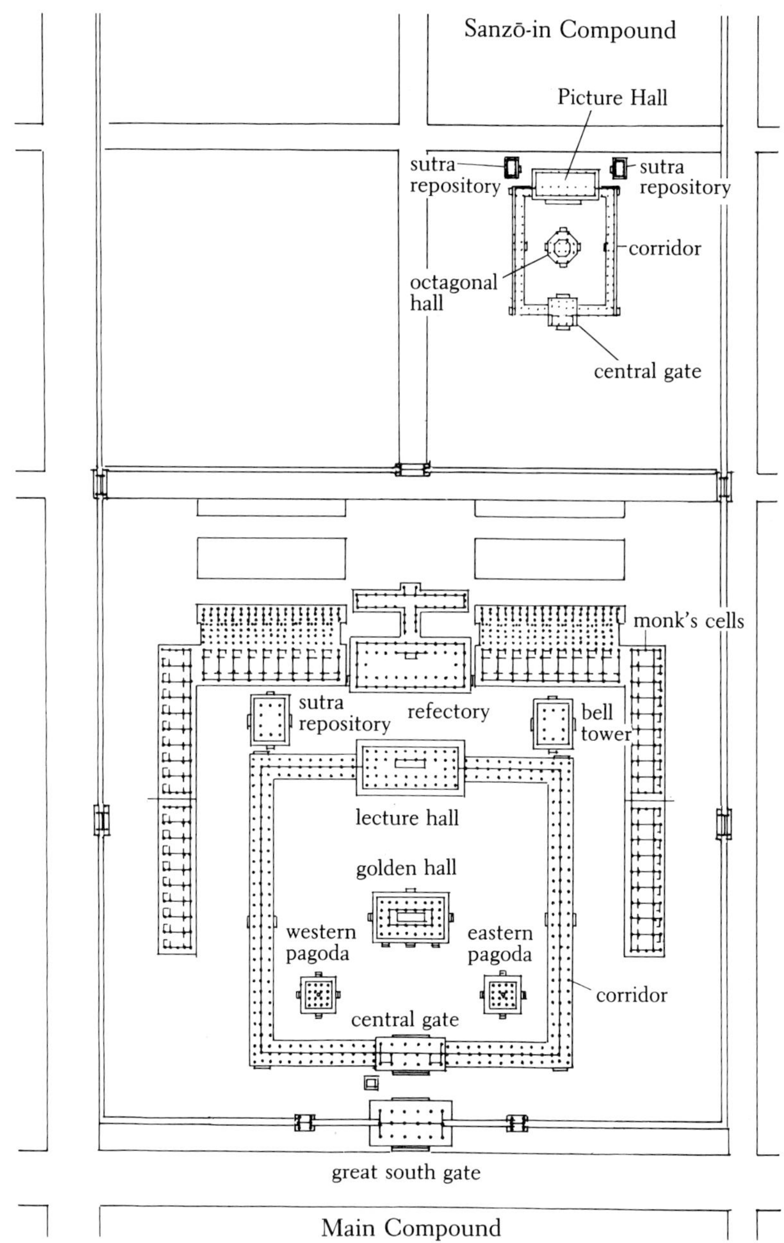

13. Layout of Yakushiji.

The capital city was moved again in 794 to present-day Kyoto, partly because the powerful Nara temples had become a political nuisance. The Yakushiji did not accompany this move, but remained in Nara. In time rival sects appeared and obtained imperial sponsorship, causing Yakushiji to decline in importance. The temple thereafter fell periodic victim to fire and natural disaster, the worst of which was the torching of the temple during the civil war of

1528. The golden statues were blackened by the flames, and every building but the eastern pagoda was destroyed. A temporary golden hall was erected in 1600 and a new lecture hall in 1852, but by the mid-twentieth century the temple had deteriorated into a scarred veteran surrounded by rice paddies and overgrown with trees, with few remaining hints of its former grandeur.

Perhaps the abbots of Yakushiji had dreamed from time to time of reconstruction. But even assuming that a craftsman could be found who was capable of such a major undertaking, there would always remain the daunting problem of funds. Before the modern era, the national government had little or no interest in the architectural reconstruction of now obscure Buddhist sects, and with the onset of the modern era, what money was available for ecclesiastical construction was poured into building and maintaining Shintō shrines as a means of legitimizing the government's claims for the emperor's divinity. Some of the newer Buddhist sects, such as Pure Land and Shingon, had grown quite wealthy on the donations from the faithful, but Yakushiji was not in such a position.

In the late 1960s, however, noticing what seemed to be a small resurgence in Buddhist faith since the end of the war, a new abbot of Yakushiji, Kōin Takada, conceived a plan to raise the funds for reconstruction. He organized a movement for copying the Heart Sutra, the major scripture of the Hossō sect, whereby worshipers visiting the temple could spend an hour or two transcribing the sutra by hand as an act of devotion, leaving a donation when they departed. Takada considered it of utmost importance that funds for the temple should come, not from corporate donations, but from ordinary people as an act of spiritual devotion; he intends it to be an eternal movement. To date, over 3,500,000 people have contributed in this way.

Because of Master Carpenter Nishioka's tremendous success with the Hōryūji restoration, he was the obvious first choice for director of the Yakushiji project. He began by making a model of the western pagoda under the sponsorship of a railroad company which was building a historical museum within the confines of a train station in Nara. Then, in 1970, Nishioka collaborated with the noted architectural scholars Hirotarō Ōta and Kiyoshi Asano on the conceptual design of the golden hall, and produced a preliminary model. Later that year, Nishioka traveled to Taiwan for the first time to look at *hinoki*.

Since the statues to be housed in the golden hall have been designated National Treasures, stringent regulations govern their care. One of these stipulates that they be housed in a fireproof structure. The design of the new golden hall, therefore, features a central core of reinforced concrete, equipped with well-concealed automatic fire shutters and sprinklers. This core is ingeniously mated to the enveloping traditional wooden structure that forms the visible edifice. The approach is quite novel, faithful to the original, and, fortunately, entirely unobtrusive. The groundbreaking ceremony was held in May of 1971, that for the column raising in April 1973, and that for the raising of the ridge beam in December of the same year. The golden hall was completed in July 1975 and dedicated in April of the following year. It had taken a total of six years from conceptual design to dedication.

The reconstruction of the western pagoda commenced in earnest in June of 1976 when Nishioka and his assistants began measurements of the 1,300-year-old eastern pagoda to determine every detail of its construction. The foundation stones of the western pagoda, which remained intact from the time the superstructure was destroyed in the sixteenth century, were excavated by the Nara National Research Institute of Cultural Properties. Nishioka's design of the new western pagoda adhered as faithfully to the original as possible, although some minor changes in concealed structural members were necessary in order to compensate for the characteristics of the Taiwanese *hinoki*.

Another consideration was the uppermost roof of the new pagoda. The roof of the older eastern pagoda, which served as a model, was not the original one, but the result of repairs in the Edo period, when it was altered to suit current preferences. For the new pagoda it was decided to recreate the original profile. In order to do this with utmost accuracy, Nishioka calculated the ultimate shrinkage of the lumber in the entire structure and the degree of settling it would cause, keeping in mind that the vertical and nonloadbearing central pillar would not settle at all. Thus, when the central pillar was installed, it was raised on wooden wedges a few centimeters above the foundation stone, so that the wedges could be withdrawn incrementally over a period of years as the surrounding structure settled around it. After several years, the wood had fully stabilized, the roof angle was correct, and there were no gaps where the central pillar emerged from the roof. In October of 1977,

14. The central gate of the main compound as reconstructed (completed 1984).

the groundbreaking was held for the new pagoda; relics of the Buddha, obtained from Gandhara in Pakistan, were enshrined under the central pillar in May 1978; the pagoda was completed in October 1980.

Next to be rebuilt was the central gate (fig. 14), with construction beginning in 1982 and ending in 1984. Though considerably smaller than either the pagodas or the golden hall, it is still quite large by contemporary wooden construction standards. Each side of the gate contains a niche for a guardian statue. These sculptures, standing over five meters in height and originally made of clay, have been reproduced in wood by Kannya Tsujimoto, a renowned carver and restorer of antique Japanese sculpture.

Prior to the completion of the central gate, Nishioka had produced basic designs for the remainder of the main complex, including the surrounding corridor (currently scheduled to be begun in 1990), a new lecture hall, refectory, bell tower, and sutra repository. The entire reconstruction project, barring any major interruptions, is expected to be completed in the year 2030. Since Nishioka is now (in 1988) eighty years old, he has concentrated on finishing the basic design work for the entire project and on ensuring that his apprentices are adequately trained to complete the project when he is no longer able to participate. The work will thus continue posthumously in his name.

The ultimate aim of the Yakushiji project, however, is not restricted to the historical restoration of the temple to its eighth-century appearance; the final goal is to upgrade Yakushiji to the status of head temple of the Hossō sect, a position held by a Chinese temple until the Communist Revolution. The relics of one of the sect's founders, the Chinese monk Hsüan-tsang (Genjō Sanzō), were brought to Japan in the 1940s. In 1979, Nishioka, Professor Ōta, and Abbot Takada took the first step toward creation of a new complex to house these relics. The construction of this complex, the Sanzō-in, was decided upon at the end of 1979 (prior to the completion of the new pagoda). In 1981 Nishioka once more traveled to Taiwan to select the trees.

The Sanzō-in (see fig. 13) reproduces many of the features of the main complex in miniature. There is a central gate, a surrounding corridor, and, in the center, the most important structure, an octagonal hall called the Genjō-dō, which will house the founder's relics under a life-sized wood and lacquer image. The other major building, which will be discussed in detail below, is of a type known as a "picture hall" (*eden*), so named because the interior walls, and sometimes the ceiling, are decorated with educational or spiritually evocative paintings. In this instance, the pictures are to depict scenes from the Silk Road, a trade route that indirectly linked Japan with the Middle East through China and India and played a vital role in the flowering of Japanese culture during the sixth, seventh, and eighth centuries. Hsüan-tsang himself had passed over this road in order to bring back sutras from India for the monasteries of China, and so it was deemed fitting that his picture hall should celebrate his travels. The ceiling will be painted with a design of the heavens.

The construction of the Picture Hall was begun following the groundbreaking ceremony for the Sanzō-in in November 1984, and completed in 1987 (except for the paintings). The octagonal hall was completed in late 1987, and the central gate in early 1988. The surrounding corridor and a pair of sutra repositories (housing one million small wooden towers, each containing a copy of the Heart Sutra made by a worshiper) are scheduled to be completed in 1989. The entire Sanzō-in complex is to be dedicated in 1990, whereupon work will recommence on the main temple complex.

TEMPLE DESIGN: The Sanzō-in

The design of Buddhist temples in China and Japan is a complex product of slow evolutionary change. Indeed, the resulting welter of cross-influences, hybridizations, alterations, and idiosyncrasies has been a source of endless scholarly debate. Despite this diversity, however, there has always been a tendency toward codification and standardization, particularly with regard to scale and proportion. In China, building design was first codified in the twelfth century with the publication of government building standards, superseded by a second important manual in the eighteenth century; Japanese carpentry manuals were similarly standardized in the Edo period (1603–1867). Carpenters throughout Asian history kept individual records as well, many of which have survived. Perhaps more important than written records, however, was the fact that an existing building's proportions are for the most part readily apparent to the naked eye, subject only to limitations inherent in the perception and analytical abilities of the observer. In other words, once the image of a building is firmly fixed in an expert carpenter's mind, he can, with some experimentation, reproduce it. Thus for a carpenter like Nishioka, notations and measurements (important as they are) take second place to the mental images which inform his ideal of beauty.

The sense of beauty is to some degree culture-specific, and it took some two hundred years of experiment and refinement for the Japanese to reshape borrowed Chinese aesthetic forms into configurations closer to their own ideals. Roof curves became more gentle, certain structural elements more delicate, and composition at times asymmetrical. By the time Yakushiji was built, the initial adaptive phase was almost complete. By the Kamakura period, beginning in the late twelfth century, a simple, economical style had been perfected that could be considered completely Japanese, the *wayō* style.

Most temple carpentry today is historical reconstruction—as opposed to original design—which makes it closely resemble detective investigation. This is true of the main compound of Yakushiji. Study of other temple compounds, such as Hōryūji, shows that each was intended as a coherent aesthetic entity, held together by shared details and proportional relationships. The master carpenters of ancient times tended to apply the same formulae to all the buildings in a particular temple compound, but varied them more or less for other temples. Thus, temple compounds of roughly the same period show a high degree of individuality while retaining a certain generic kinship. The key to design reconstruction is to discover the original proportional formulae, and to reapply them harmoniously to several structures of widely differing sizes and plans. Proportional formulae generally fall into two categories. The first concerns the proportions of the site; the second, the proportions of the building's components themselves. These two systems are not, of course, independent; the overall layout of a carefully proportioned temple complex must be based on the overall sizes of its major buildings, which in turn are modified by their own internal proportional relationships. Archaeological and historical research is an invaluable aid in uncovering such formulae.

One of the primary archaeological records are the original foundations, with the size and location of column base stones particularly reliable clues to the overall size of the building. Further valuable hints can often be found in pictorial and written records. At Yakushiji, the existing eastern pagoda provided almost all the information necessary to rebuild not only the western pagoda but also the other buildings in the main compound. Such information included data on relations between column height and width, between column height and the size of bracket complexes, and between the spacing of main floors and "false" roofs, as well as such details as curve shapes and decoration. Early investigations showed that the temple layout as a whole was governed by a simple proportional system based on the height of a single pagoda (fig. 15). This height, 120 *shaku* (the *shaku* is a Japanese standard of measurement that varies from era to era and in different parts of the country but is roughly equivalent to one foot; fig. 16), dictates the spacing of the pagodas themselves, the dimensions of the

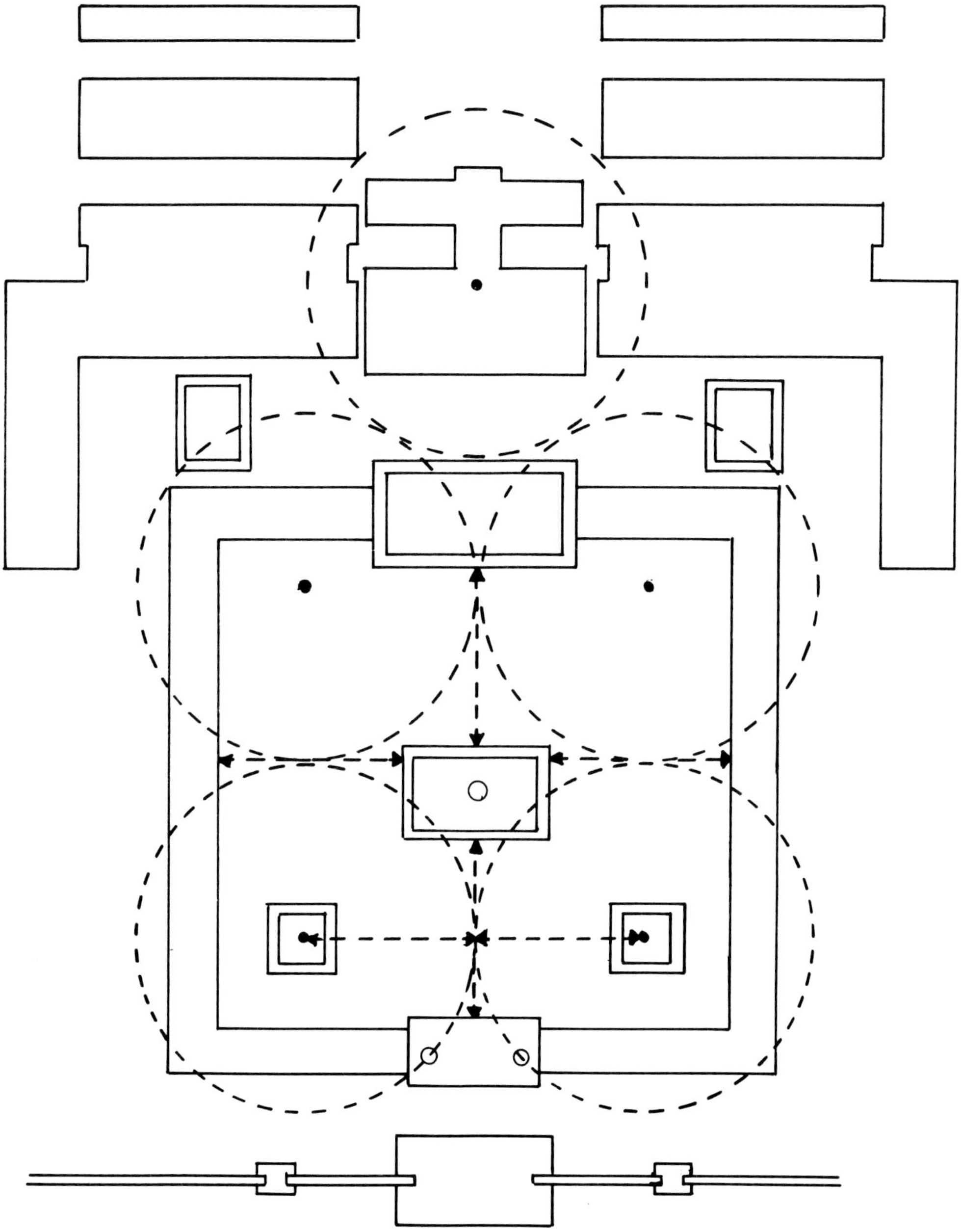

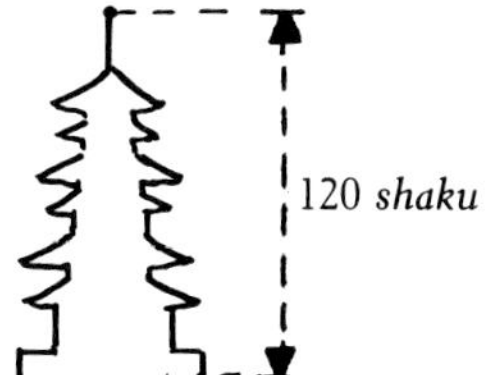

15. Proportional diagram of Yakushiji main complex. The circles indicate major determinants: that is, 120 *shaku* or Japanese "feet" (the distance from foundation to finial of the eastern pagoda).

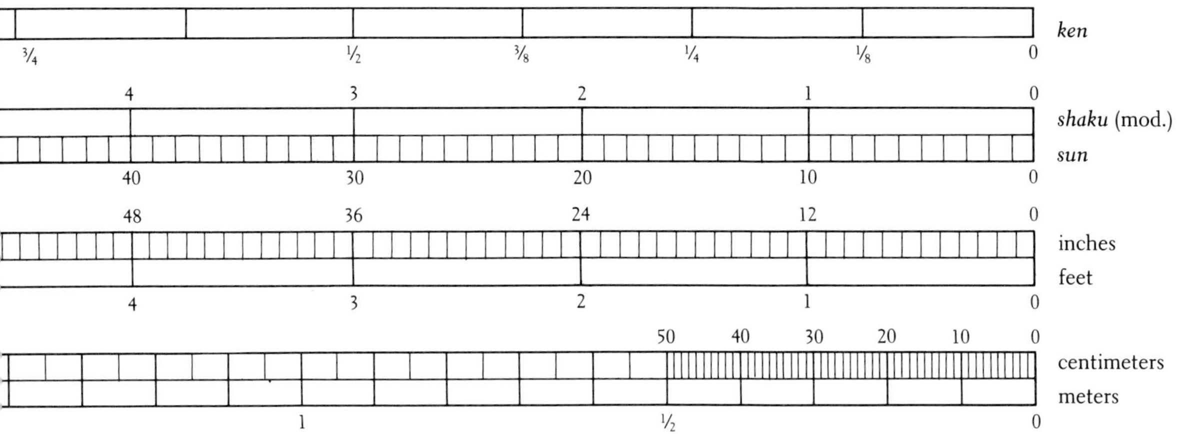

16. Relative scale of measurements. (The Nara-period *shaku* is 0.98 of the modern *shaku*).

surrounding corridor, and the placement of the other buildings within the compound. The interrelationship among the parts of a temple compound is such that very little information is necessary to reconstruct the layout of the whole, much as a paleontologist might reconstruct an entire dinosaur from a few bones.

Because the Sanzō-in is a new complex (figs. 17–18, color pl. 2), the designers thought it unnecessary to follow the eighth-century style of Yakushiji's main complex. Rather, they viewed this as an opportunity to demonstrate construction in a later historical style, and settled on the Kamakura-period *wayō* style because of its relative economy and quintessentially "Japanese" nature.

Thus Nishioka and Ōta produced for the Sanzō-in compound an original design based on several historical examples, incorporating features drawn from all while maintaining consistent proportional relationships. The overall size of the complex was determined by several factors. First, the religious function of the buildings dictated the number of people to be accommodated, and therefore the approximate size. Next, available land and the projected budget were considered. The result is the largest possible group of buildings that can be built on the given site for the designated amount of money, keeping in mind the specific religious and technical design requirements. Some factors, such as the south-facing orientation, are dictated by Buddhist canon. Additionally, the buildings had to meet the various requirements of ceremonies, processions, and the like. These aspects all have traditional solutions and extant models.

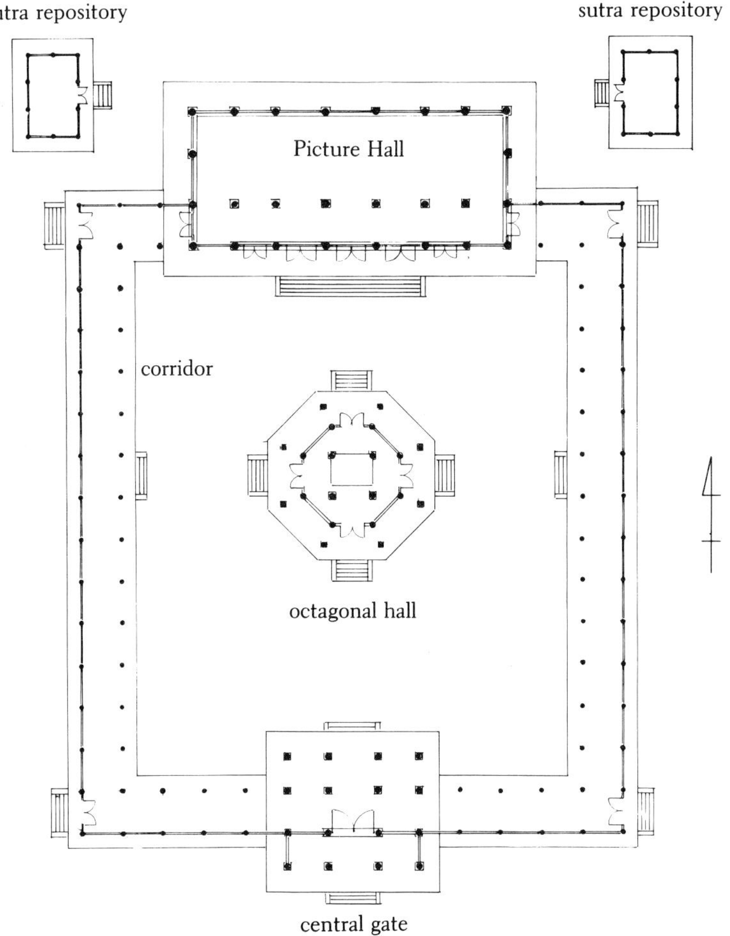

17. Plan of the Sanzō-in.

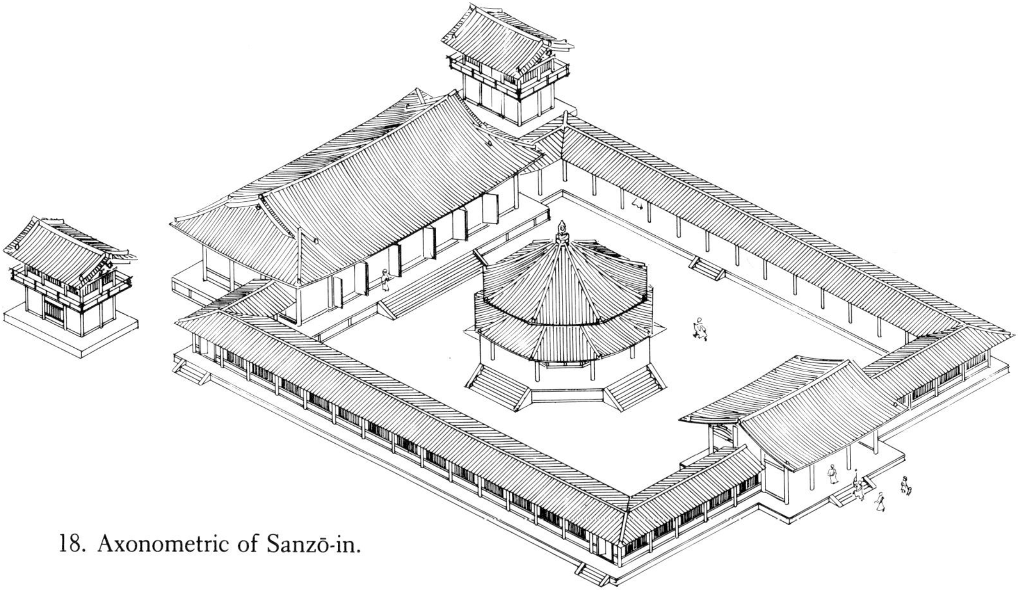

18. Axonometric of Sanzō-in.

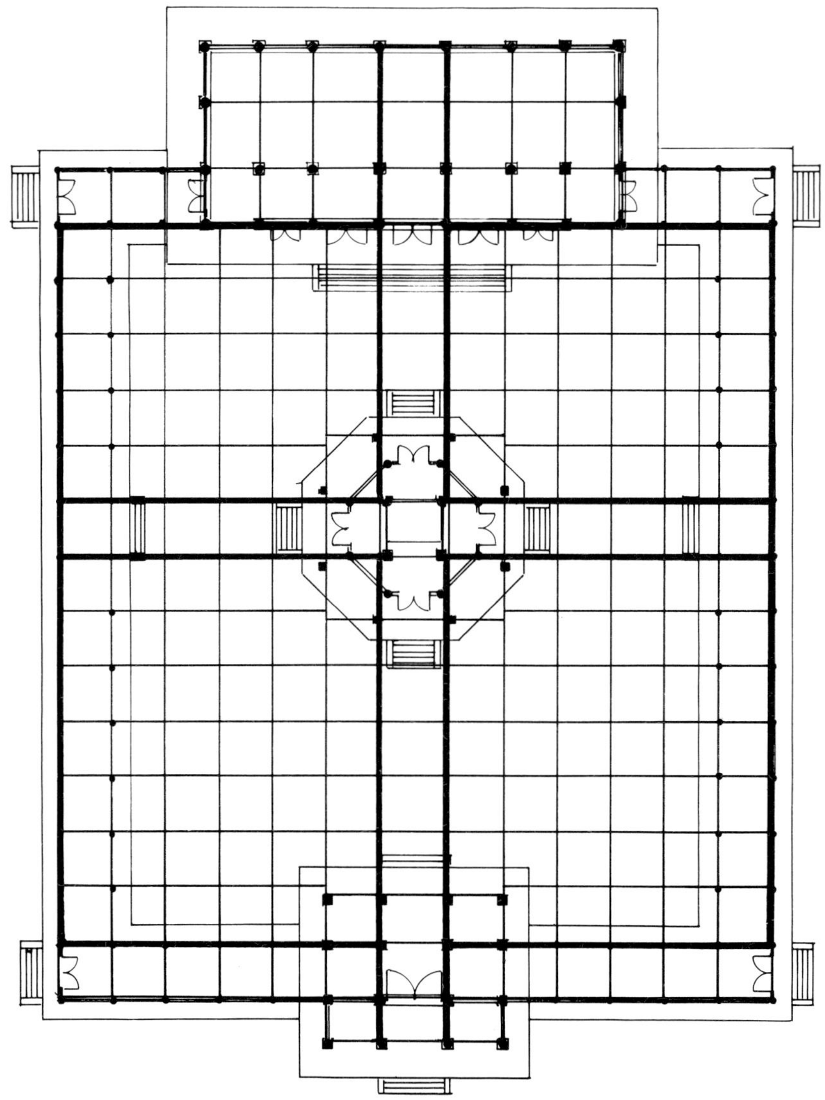

19. Sanzō-in proportional diagram: major axes of the compound are plotted on an imaginary grid defined by column spacing. Grid module is roughly ten *shaku*, but varies where necessary.

After the overall layout was decided, the proportional formula was developed, and Nishioka began to apply it to the actual design, which underwent several revisions as details were established (figs. 19–23). The Picture Hall was the first building to be erected and the first to be designed in detail. This stage of design work culminated in a set of blueprints not unlike those used in any modern construction. The actual work itself was done in collaboration with a major con-

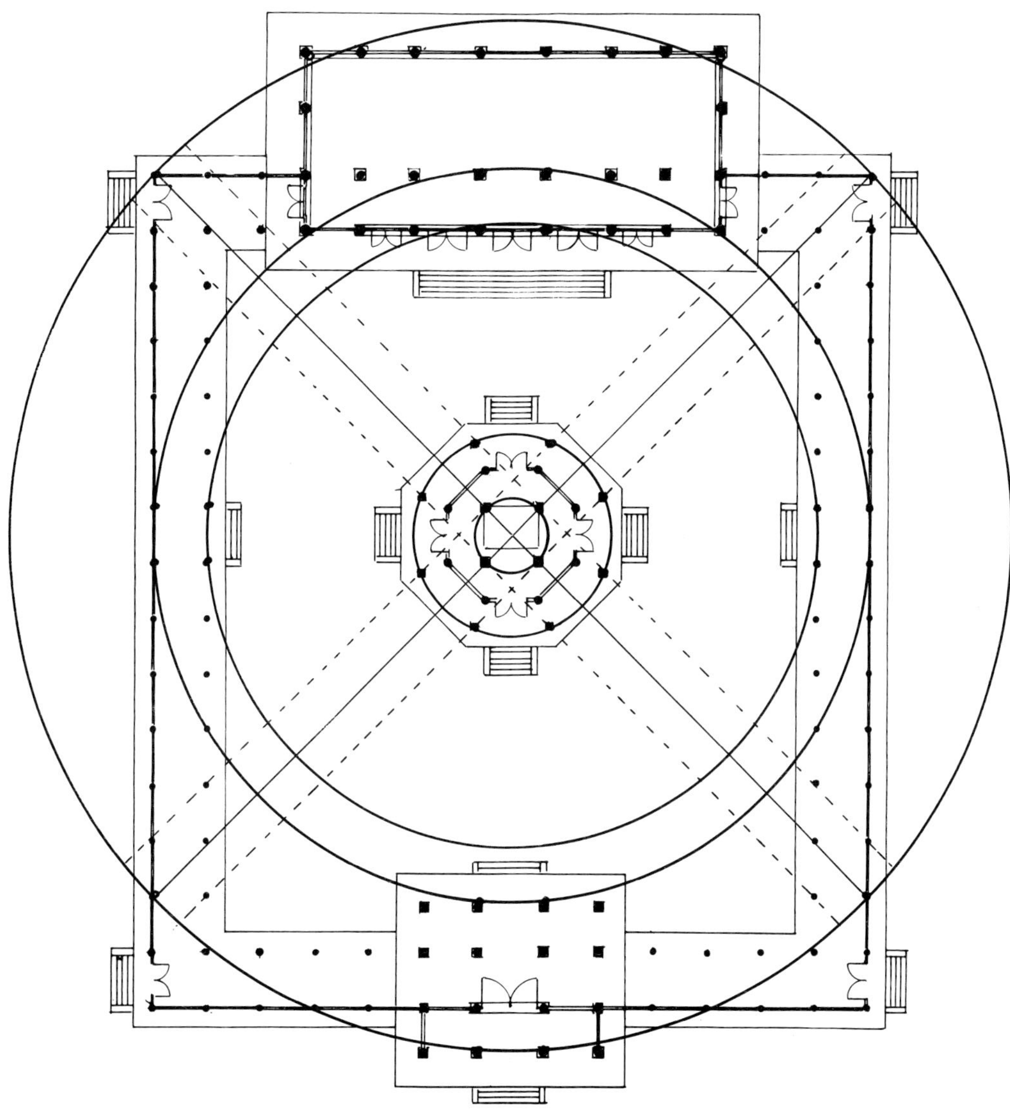

20. Sanzō-in proportional diagram: major dimensions plotted as circles centered on the relics in the octagonal hall.

struction company, Ikeda Kensetsu, which provided heavy equipment, logistical support, and carried many of the carpenters on its payroll, though all were under the master carpenter's direction. It prepared and executed plans for reinforced concrete foundations, drainage, and temporary sheds within which carpenters could work unhampered by inclement weather. After a special groundbreaking ceremony held in November 1984, the laying of the foundation began.

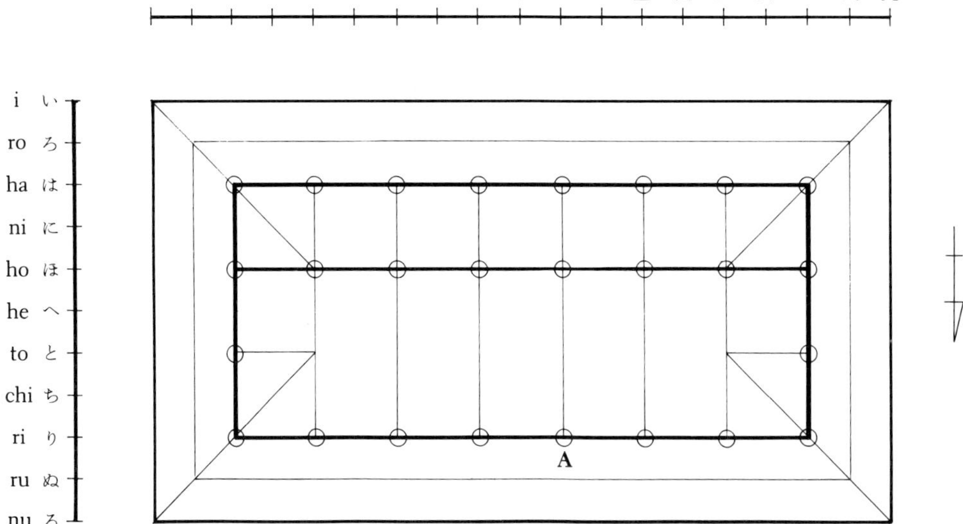

21. For convenience in labeling parts, carpenters use a grid system based on traditional numbers and syllables. A simplified version is shown. For example, the column marked A in the illustration is referred to as り十一, which is read *ri jūichi* ("ri eleven"). See also figure 95.

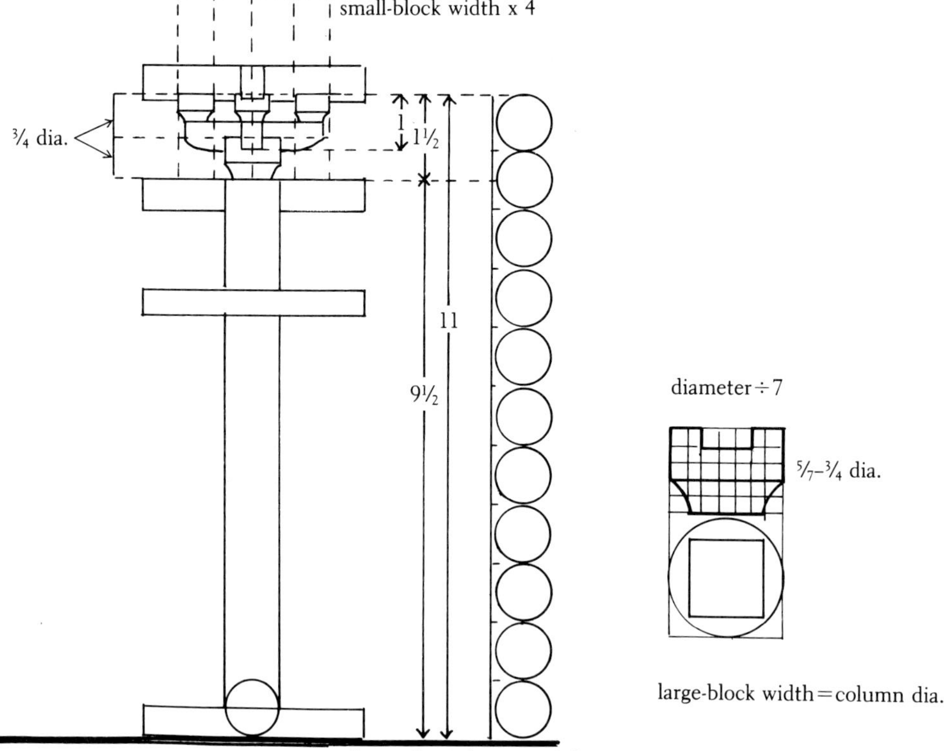

22. Proportions of column diameter and height, eaves, etc.

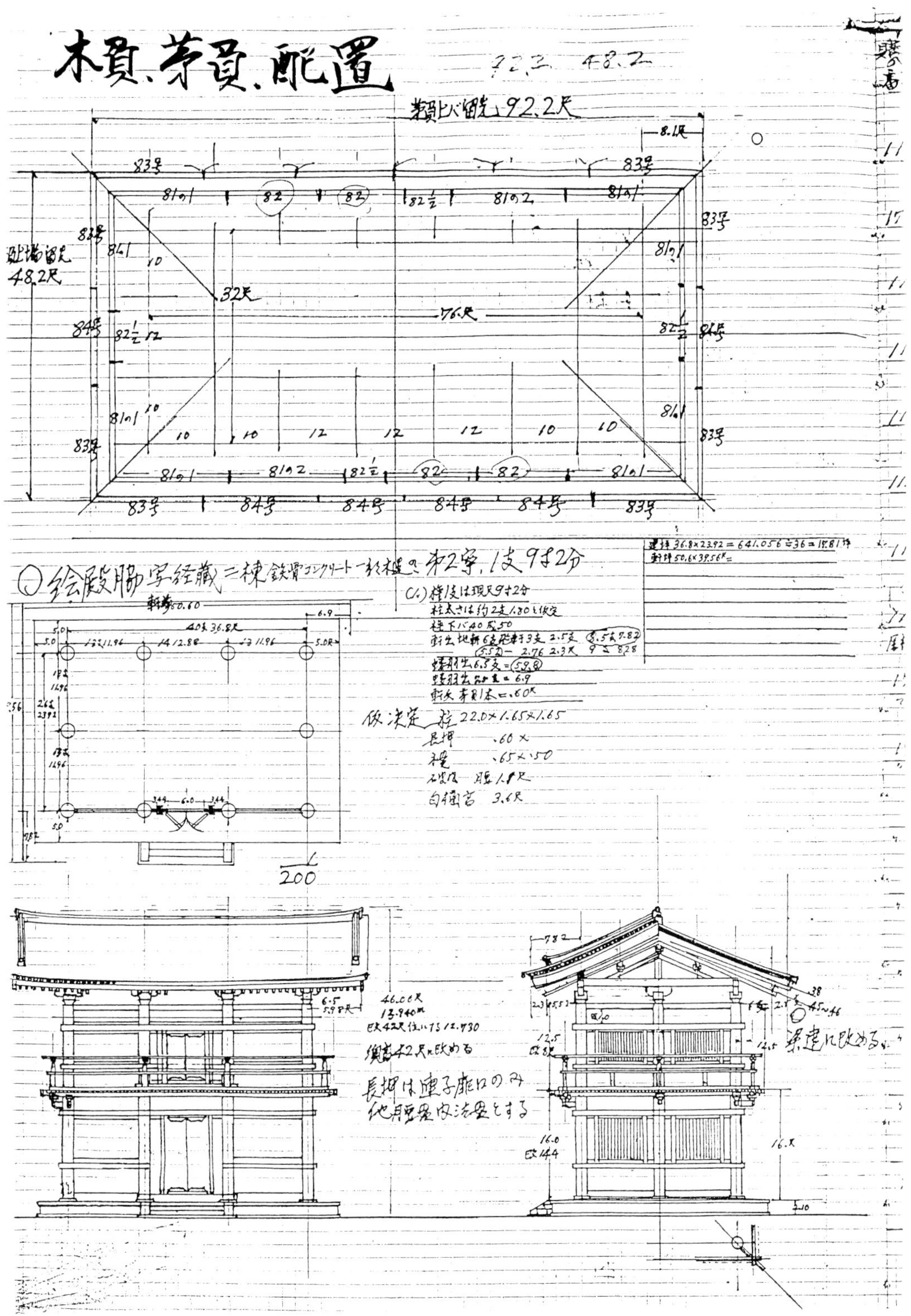

23. A page from Nishioka's notebook dealing with dimensions and proportions of sutra repository and Picture Hall.

24. Templates used in fabricating parts (here for the octagonal hall).

At this point, Nishioka's drawings were passed on to his chief assistant. Although it is conceivably possible to make the parts directly from printed plans, temple carpenters traditionally use extremely precise full-size wooden templates (fig. 24). The fabrication of these is the chief assistant's job, and requires the ability to visualize complex three-dimensional intersections and to project them onto two-dimensional surfaces. Every detail of every part must be thought out at this stage and recorded on the templates.

Many parts have compound curves; that is, they curve in both plan and elevation. These curves must be accurately visualized and described at this stage. In addition, when such curved parts intersect with others (often with two or three at the same point), the joints must be indicated in detail on the templates. And lastly, information pertaining to true horizontal and the ultimate height of the part from ground level—constantly checked during construction—must be included.

This work is performed in a special room with a smooth floor onto which the plywood sheets are tacked. The chief assistant lays out the various elements in pencil and sumi ink, with the help of squares, inking lines, and other wooden templates. During this process, the relationships among the separate parts are continually checked by placing one sheet over another like drawings. When the layout is perfect, the templates are meticulously cut out by hand and delivered to the assistant master carpenters. These templates are preserved for use when repairs become necessary in the future.

SELECTING THE WOOD: Rules of Thumb

During the ages when timber was in ample supply, a Japanese carpenter could be assured of adequate prime building material close to home for all but the largest projects. Over the centuries several rules of thumb evolved concerning the optimum use of wood which reflect the value placed on continuity between the wood's natural, living state—what Nishioka referred to as utilizing the "personality" of the wood—and its ultimate disposition within the erected framework. Although the poor state of forest resources in Japan today has made strict observance of these principles impossible, they remain as ideals.

Many of the carpenter's rules of thumb are based on geomancy, derived in large part from Chinese principles. Geomancy seeks to identify the various natural forces acting on a building site and to compensate for any perceived imbalances. For instance, according to Nishioka, the ideal building site has a mountain to the north, a river to the east, a pond or lake to the south, and a straight road to the west. When I asked how valid he thought these principles were, he remarked simply: "The Hōryūji site conforms to these rules perfectly, and the temple is almost entirely intact. Yakushiji's site has no mountain to the north and no pond to the south, and only one building—a pagoda—has survived." The evidence apparently spoke for itself. Deficiencies in the site, however, can be compensated for. If the needed river is missing, seven willow trees (a tree usually found close to rivers) should be planted in the specified location. In place of a pond, seven paulownia trees may substitute; for a road, seven plum trees; and for a mountain, seven *enju* trees, a variety of legume.

Carpenters extend the geomantic principles to human relations as well: a certain balance of personalities is required among crew members just as among pieces of timber if a project is to result in the creation of unity. "The carpenters," says Nishioka, "should all know what will be expected of them, what the goals of the project are; in order to do this, they should try to become one with the master's heart."

ORIENTATION: Depending upon its position on a hillside, a tree will be exposed to different conditions of sun, wind, moisture, and so on. Consequently, care is taken to maintain the compass orientation of the timber in the finished building, particularly in the case of columns (figs. 25–26). For instance, trees that received strong sunlight due to their location on a southern slope are best used as columns on the southern face of the building. A tree's structure is naturally designed to resist gravity, being thicker and denser at the base and thinner and lighter at the crown. This vertical relationship should also be maintained, base end always down, crown end always up. In addition, where possible, horizontal members should be joined crown-to-crown (fig. 27). This is true of inclined members as well.

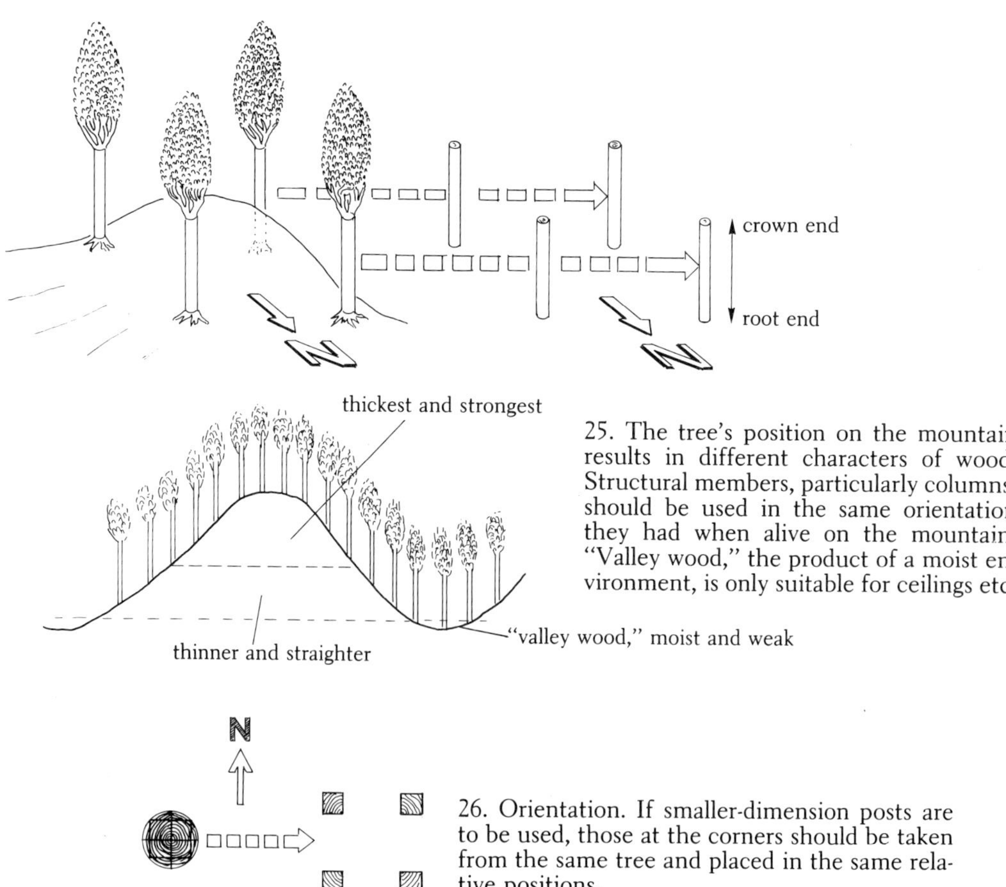

25. The tree's position on the mountain results in different characters of wood. Structural members, particularly columns, should be used in the same orientation they had when alive on the mountain. "Valley wood," the product of a moist environment, is only suitable for ceilings etc.

26. Orientation. If smaller-dimension posts are to be used, those at the corners should be taken from the same tree and placed in the same relative positions.

A tree's location while growing affects its usable characteristics in other ways as well. According to Nishioka, the strongest trees grow above the midpoint of a mountain, where they receive the best sunlight and air circulation. The trunks are thicker since the trees there are able to branch quickly and fully and develop the necessary girth to support a large amount of leaves or needles without excessive competition from neighbors. These trees should be used for columns. On the other hand, trees that stand below the midpoint, where they must compete for sunlight with trees above and thus send their trunks higher before branching, are a source of thinner logs that are relatively free of knots. These logs should be used for exposed members requiring a fine surface. Trees growing in valleys produce inferior lumber of high moisture content, but are nonetheless usable for ceiling boards and other parts requiring neither strength nor fine finish.

GRAIN AS STRUCTURE: Lumber with a curving grain can be used in ways that take advantage of its structural characteristics (fig. 28).

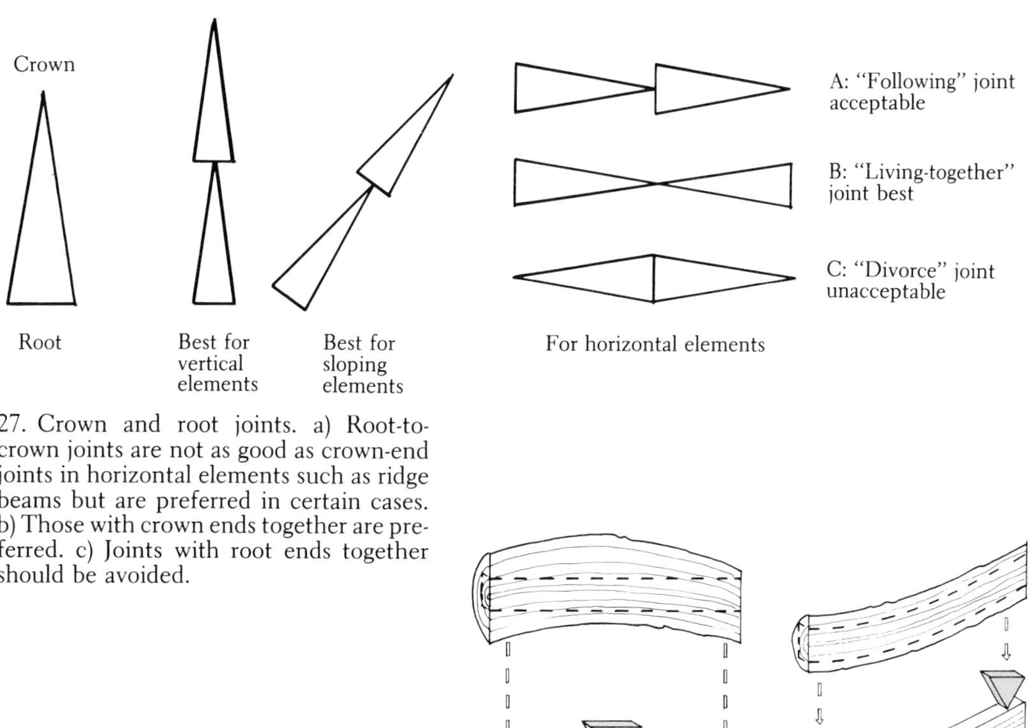

27. Crown and root joints. a) Root-to-crown joints are not as good as crown-end joints in horizontal elements such as ridge beams but are preferred in certain cases. b) Those with crown ends together are preferred. c) Joints with root ends together should be avoided.

28. On the lcft, the grain acts as a natural arch, resisting the center load. On the right, the grain is an asset when the member bears a cantilever load.

MOVEMENT AND SHRINKAGE: All wood changes shape in response to changes in temperature and moisture. The greatest movement occurs perpendicular to the grain when viewed from the end, that is, radially. Lumber should be milled so as to minimize shrinkage-related distortions, and to insure that the changes that do occur are predictable and can be accounted for in design (fig. 29). Logs with twisting grain should be used in opposing pairs so that their movements counteract each other.

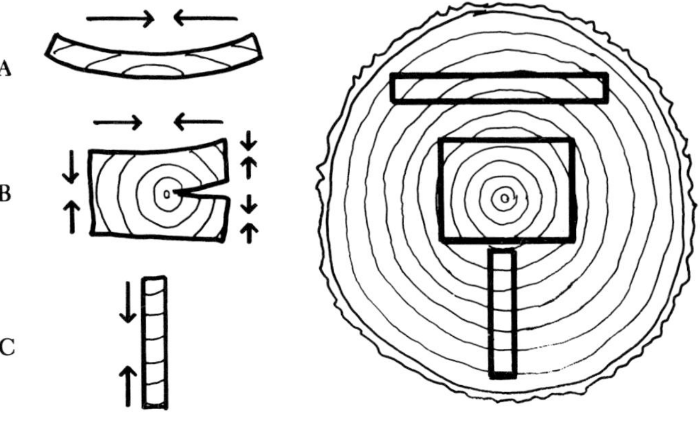

29. Arrows show direction of shrinkage. C is the most stable dimensionally.

FABRICATING THE PARTS

The main workshop for this project is a vast metal-roofed structure not far from the main Yakushiji complex and adjacent to the Sanzō-in. It is dirt-floored and warmly lit by sunlight filtering down onto the workers from a dozen or so translucent panels (fig. 30, color pl. 3). There is, of course, electricity for lights and power tools, and large doors on all sides admit additional light as well as timber-laden trucks and forklifts. One tiny room serves as a rest area, with large tea decanters, shelves for personal effects, and, in winter, a kerosene stove. In summer, there is no respite from the heat. A nearby shed of similar dimensions functions as a painting room and storage for finished parts. A third large building is jammed with enormous logs undergoing a three-year seasoning (fig. 31). Building materials are constantly being shifted about from one shed to another, ultimately finding their way to the construction site and into their predestined locations in the finished structure, marking the end of a process that began several centuries and thousands of miles away when the logs were seedlings on a Taiwanese hillside.

Activity in the workshop is never frenzied, but progresses at a steady and deliberate pace on several elements at once, rarely falling behind schedule. If a deadline seems to be approaching too quickly, extra hours are added to the work-week, an option considered infinitely preferable to rushing the work.

The actual fabrication of an element begins after the completion of its template, which the chief assistant then passes on to one of the assistant masters. The logs to be used for major components such as columns and the huge hip rafters have been selected and marked in advance; the lumber for other particular components is chosen at this point, the first stage in a process known as *ki-dori*, or wood prepa-

30. Fabricating ridge beams in the main workshop.

31. Taiwanese *hinoki* logs stacked for seasoning. The one in the foreground is nearly two meters in diameter.

32. Rough-cutting material on industrial-grade woodworking machinery.

ration. Here the carpenter must examine such characteristics as wood grain and knots, and try to predict how the wood will change. He must visualize the finished components that lie dormant within the rough lumber, and do so in such a way as to take full advantage of each log's idiosyncrasies. Many carpenters consider this to be the single most important stage of their work; the skills it requires are not physical but mental: imagination, visualization, an intuitive grasp of the wood's "personality."

Ki-dori proceeds in several stages, as each log is reduced to smaller and smaller components. The primary milling is conducted at a nearby sawmill, and the lumber transported back to the workshop, where the smaller pieces are roughly cut on large-capacity band saws and table saws (fig. 32). Since most components come in sets, usually several pieces of rough lumber identical in size and configuration need to be cut. They are then moved to a corner of the shop.

The workshop is spacious enough for about eight different major operations to be carried out simultaneously. Generally speaking, one assistant master undertakes the fabrication of an entire set of identical components, assisted by a junior apprentice from time to time. The carpenters confer with each other frequently, but interference in another's project is taboo. Each takes full responsibility not only for his own work but that of his helper as well.

33. Kogaitō-san, an assistant master carpenter, does initial layout with the aid of templates, steel square, inkpot and line, and bamboo marking pen.

After basic preparation, the information from the templates is transferred to the pieces to be cut. This is called *sumitsuke*, or layout (fig. 33). Using the templates and shop drawings as a guide, cut-lines and other notations are transferred to the workpiece in ink. The primary layout tools include the inkpot and line, metal square, and bamboo marking pen. Marks are made in black ink, with occasional notations in red. The ink is indelible, so any corrections require the wood surface to be planed off, excising previous notations. Generally speaking, identical parts are all laid out at once, but this is a matter of personal choice. The indelibility of the ink is an asset for future restorers, who frequently refer to the original labels and notations; disassembly and reassembly, as well as tracing the original function of a piece that has subsequently been moved, are all facilitated by adequate notation.

34. An apprentice chisels out notches in a ridge beam previously laid out by an assistant master carpenter.

Initial shaping is now usually done with power tools, but it can be carried out by hand. Use of power tools is discouraged until the apprentice has mastered the requisite hand tools; the aim is to provide the carpenter with as many options as possible. It is not uncommon to see a carpenter laboriously cutting or planing a board by hand when an electric tool would be faster and equally accurate. Hand tools, particularly Japanese hand tools, provide a tremendous amount of subtle tactile information about a particular piece of wood. The delicate readings and detailed observations possible when working slowly by hand are lost with a noisy, dangerously gyroscopic blade or drill. But again, the choice of tools is personal. The apprentice sometimes finds himself the victim of conflicting suggestions, forced to remember each of his seniors' predilections in order to avoid criticism. Fundamentally, there is no operation that cannot be carried out through the judicious application of saw, hammer and chisel, and plane (fig. 34).

One remarkable fact, alluded to above, is that each carpenter works in relative isolation. His templates, plans, and knowledge constitute only a partial picture of the overall interrelationship among the parts. When instructed to provide a mortise, he provides it; when instructed to shape a compound curve, he shapes it. While it is dangerous to generalize, it is probably safe to say that a Westerner, particularly an American, often finds it difficult to perform complex work without understanding exactly how his job relates to the whole. As a result, he asks a lot of questions. There is, perhaps, an underlying belief that if one understands the overall intention, one may be able to avoid errors or even improve upon the design of one's superior. This does not seem to be the case with the Japanese carpenter. He may not always know who is making the tenon that will fit his mortise, or why the curve is necessary, but he is fully confident that if instructions are followed, the pieces will fall perfectly into place. In short, he trusts not only his superiors but his colleagues, as well as the system in which they work. This is an outgrowth of, and the reason for, the long apprenticeship.

Each piece requires several successive operations of cutting, trimming, and smoothing, with the final surface for exposed pieces demanding increasingly finer planing to achieve a glasslike smoothness (fig. 35). The finished components are then handed over to the painter.

As mentioned above, painted architecture was an innovation brought to Japan from Korea along with other techniques in the sixth century A.D. The Japanese people had up to that time constructed durable structures of unpainted wood, and in fact seem even today to prefer its natural surface, color, and grain. After painted Buddhist architecture had taken root and evolved in Japan, unpainted wooden temples began to make their appearance, and many of them have now withstood the test of centuries. I believe it is safe to say that painting is not a question of protecting the wood, and this applies particularly to *hinoki*, but represents an aesthetic choice.

Available pigments were relatively few in ancient times. Red was obtained from iron oxides, green from copper oxide, yellow from earth, white from the calcium in bones and seashells, and black from carbon. Bright blues were precious, requiring lapis lazuli, but dark blue could be obtained from indigo. In eighth-century Nara, temples were usually painted red overall, with green and white details. Ceilings often featured complicated floral patterns in a variety of colors against a white ground. These principles are followed at Yakushiji.

Pl. 1. Corner detail of the Picture Hall.

Pl. 2. Three-quarter view of the completed Picture Hall.

Pl. 3. Carpenters inside the workshop, preparing columns and horizontal brace beams.

Pl. 4. Bracket complexes being fabricated as identical sets.

Pl. 5. Intricate joinery at juncture of wall purlin and bracket complex.

Pl. 6. Bracket complexes and blocks-on-struts, with main beams awaiting placement at side.

Pl. 7. View of gable end of half-finished roof structure.

Pl. 8. View from above of almost completed roof structure.

35. Hayashi-san applies a hand plane to an outside horizontal brace beam. The shavings are tissue-paper thin and several meters long.

The painting is usually carried out by one man (perhaps with a single assistant), who begins by mixing the colors from natural dry pigments, water, and a seaweed-derived binder. He then coats the areas of the individual components that are to be exposed, leaving the other surfaces bare. A few days are generally allowed for the paint to dry completely before another coat is added. Four coats are applied in all, and then the pieces are stacked to await final assembly.

SECRETS OF ENDURING CARPENTRY

Temple components are relatively large, but their dimensional tolerances are very fine, particularly in the case of surfaces that will remain visible. Allowances must be made for wood movement: expansions and contractions due to changes in temperature and humidity, gradual deflections under load. Cumulative shrinkage from moisture loss in the wood's inner cells must be considered. For all these reasons, craftsmen speak of wood as alive and breathing.

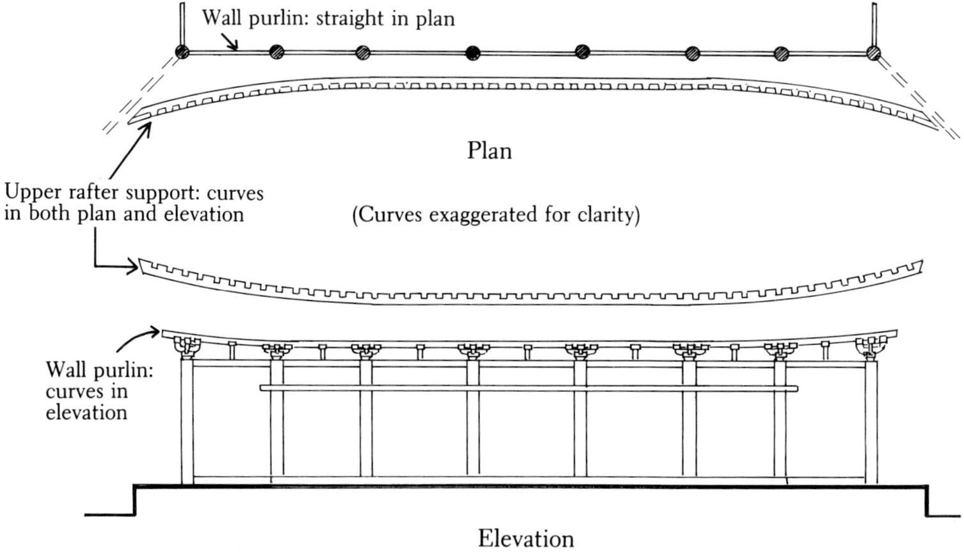

36. Curves achieved without bending the wood. The gradual transition from the horizontal tie beam to the graceful upward sweep of the eaves occurs in several stages. The first compensation is made at the wall purlin, which is about four centimeters higher at the ends than in the center; the upper surfaces of the corner bracket sets are slightly beveled in order to accommodate this curve (see figs. 110, 119). The wall purlins are not bent, but shaped to the proper curve (see also figs. 121–22). The upper rafter support and the eaves support are composed of linked segments that form a more pronounced compound curve. This curve, evident in both plan and elevation, is also formed without bending the wood. See also figures 149, 156.

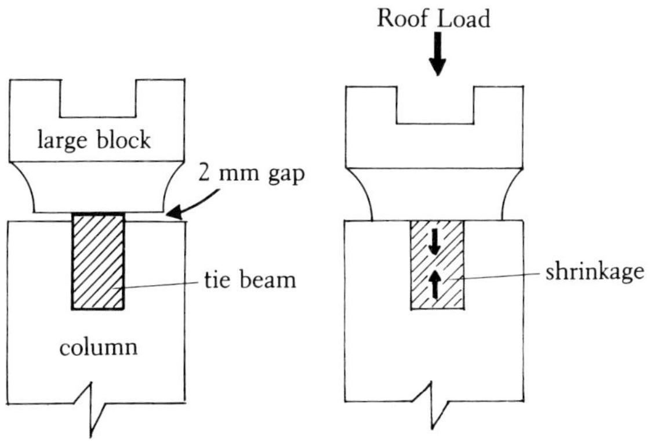

37. Upper column area. Here we we see another typical example of compensation. The columns and tie beams, which must eventually lie flat and evenly support the large block above, are constructed with a two-millimeter difference in height. The tie beam will shrink and be compressed by the roof load, giving the column and tie beam the desired flat surface. The left-hand illustration shows proper shrinkage allowance, that on the right the desired final configuration.

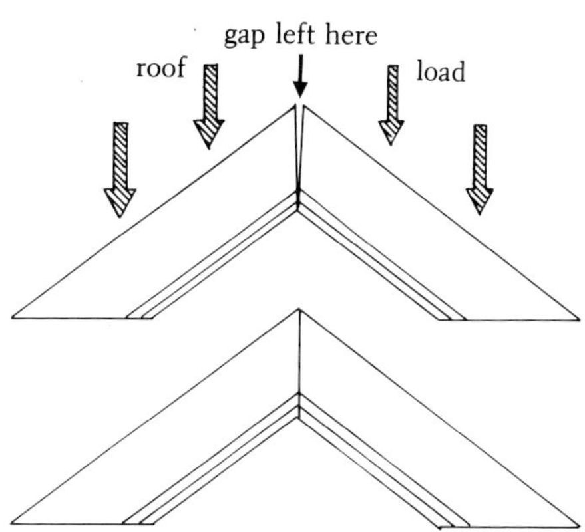

38. Similarly, a small gap left at the point where the two gable fascia meet will close as the roof settles. If this were not done, a gap would eventually appear on the bottom of the joint.

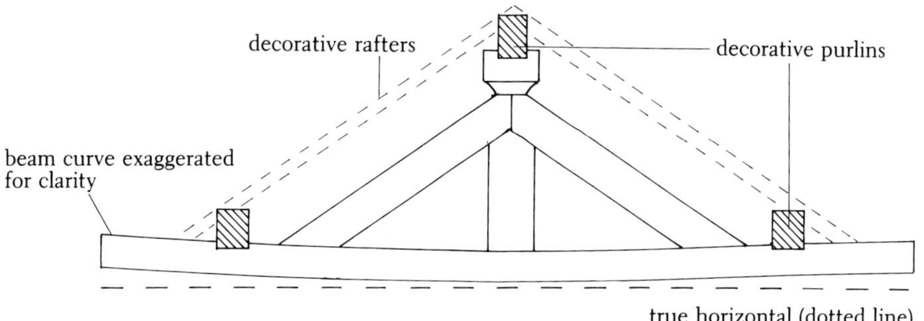

39. The gable area illustrates another type of detail. The "horizontal" beam at the base is actually curved, so that the enclosed area is not precisely triangular and therefore cannot be laid out as such. This complicates the treatment of the joints between the diagonal braces and the lower beam. See also figure 167.

One level of mastery lies in accuracy: straight lines are perfectly straight, joints snug, curves gradual and even. Another level entails the same degree of accuracy but takes wood movement into account. Will this joint still be snug a hundred years from now? Will this curve remain attractive after sagging under a century of load? Here I have given a few examples of the kinds of details that set master carpenters apart, but which often go undetected by the untrained eye.

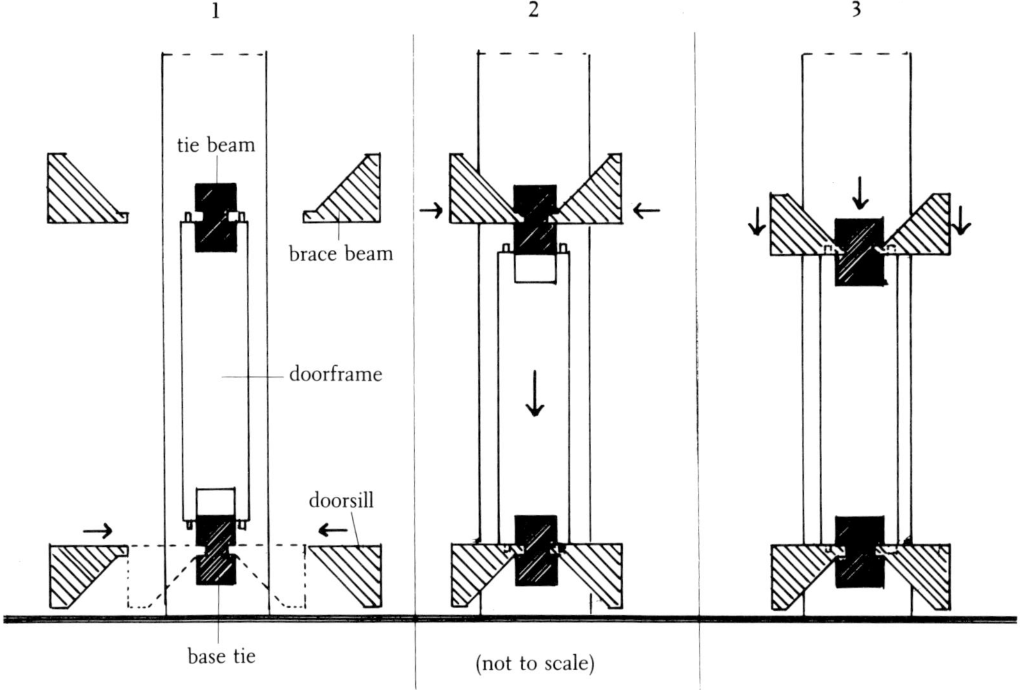

40. The doorsill, subject as it is to damage from water and visitors' feet, is usually the first part to be replaced. The entire piece is thus designed to be removable. 1) Both doorframe and upper penetrating beam are slid upward to allow the doorsill to be inserted. 2) The doorframe is slid downward, engaging the doorsill with small tenons. The upper brace beams are inserted into the tie beam. 3) The tie beam is slid down, engaging the brace beams with small tenons at the top of the doorsill. The tie beam is wedged into place. Removal procedure entails the replacement of a small section of mud wall above, a relatively simple matter. See also figures 100–101.

TOOLS

It is sometimes said that the tool is the soul of the Japanese carpenter just as the sword was the soul of the samurai. The comparison is apt in many respects. To be more specific, the tool is the means by which the carpenter leaves a trace of his spirit on the soul of the tree, the medium through which a marriage of souls is brought about. As a result, carpenter and tool serve each other, and together they serve the tree. There are regular occasions, such as at New Year's, when the carpenter expresses his gratitude for the service his tools have rendered. Offerings of sake and rice cakes are made, and the carpenter promises, despite the great abuse his tools have endured, to treat them with greater respect in the future.

During the Asuka period, when the first Japanese Buddhist temples were built, only four types of carpentry tools were in use—curved-bladed chisels, axes, adzes, and "spear planes" (*yariganna*). Today, an accomplished carpenter possesses a few dozen chisels, a dozen or so saws, a similar number of planes, a couple of axes and adzes, four or five hammers, eight or nine layout and marking tools, several sharpening stones, and assorted other tools. In addition, the modern carpenter has at his disposal a variety of power tools: saws, drills, planers, mortisers, screwdrivers, wrenches, and so on. Still, the hand tools remain essential to his work, and in all cases they reflect a preindustrial way of thinking about wood and steel carried to the extremes of refinement and subtlety.

Since the heart of a tool lies in its steel, preindustrial toolmaking, like swordmaking, necessitated a time-consuming process of lamination of hard and soft steels to create the required characteristics of rigidity and resiliency, which are not unlike the characteristics of wood itself. Often this affinity between steel and wood is emphasized in the design of the tool, where the laminations stand out in a pattern reminiscent of wood grain.

"Modern steel is no good," says Nishioka. "Swordmakers have always tried to find antique steel to use. A famous swordmaker . . . once told me that good tools could probably be made out of nails from Hōryūji. . . . I had a plane blade made from some, and used it on the reconstruction of the Hōryūji golden hall" (Tsunekazu Nishioka, *Ki ni Manabe* [Tokyo: Shōgakkan, 1988], 26–27). According to Nishioka, the ancient steels were produced from rich ore smelted at relatively low temperatures, resulting in a resilient metal that cannot be reproduced from the depleted ore available today.

Marking Tools

Steel squares, which are available in both metric and traditional Japanese measurement scales, are thinner and more flexible than those generally used in the West (fig. 41). Adjustable bevels were introduced from the West in the modern era, but modified to suit the Japanese preference for thinner, all-steel tools. The inkpot and line, similar in function to a Western chalk line, is equipped with a reservoir filled with ink-soaked cotton for inking the line and dipping a bamboo marking pen into. The line, after being secured to one end of the piece to be marked, is drawn

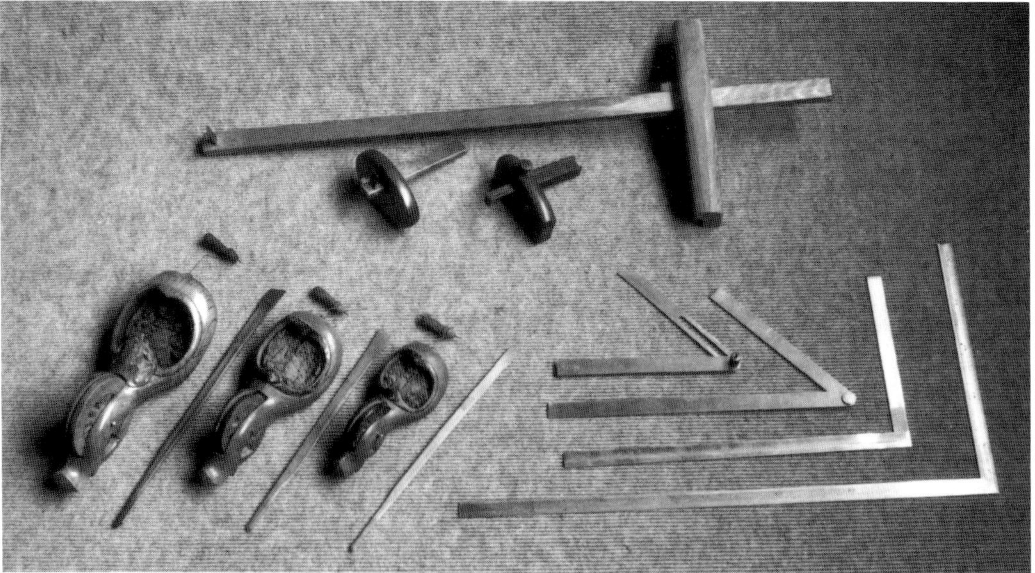

41. Layout and marking tools. Top: one large and two small marking gauges. Lower left: three inkpots with line and bamboo marking pens. Lower right: two steel adjustable bevel gauges, a small square, and a regular square.

tautly through the ink and flicked onto the wood, leaving a crisp line. These inkpots can be very ornate, sometimes carved with a miniature turtle (a symbol of longevity). Traditionally made of wood, the more common versions are now of plastic. The most frequently used ink colors are black and red. Adjustable marking gauges, available in many sizes, are used to inscribe lines at a set distance from the edge of a board. Some versions can scribe two or more lines. All consist of a rounded piece that fits in the palm and is slid along the edge of the piece to be marked. Through this rounded piece slides a wooden arm with a small steel scribing blade at one end. The arm is held in place by a small wooden wedge that is adjusted by tapping with a mallet.

Axes and Adzes

Axes and adzes, which come in several sizes, are used for quickly removing large quantities of wood; in trained hands they can do extremely fine work (fig. 42). All have a wooden handle fitted to a socket in the laminated steel blade. Proper operation of the adze requires a handle that is curved to fit perfectly the individual user's body and places the cutting edge at the optimum angle. Bending the adze stock to the proper angle is a small art in itself, accomplished with the help of water and sunlight. A perfectly curved stock assures flawless performance.

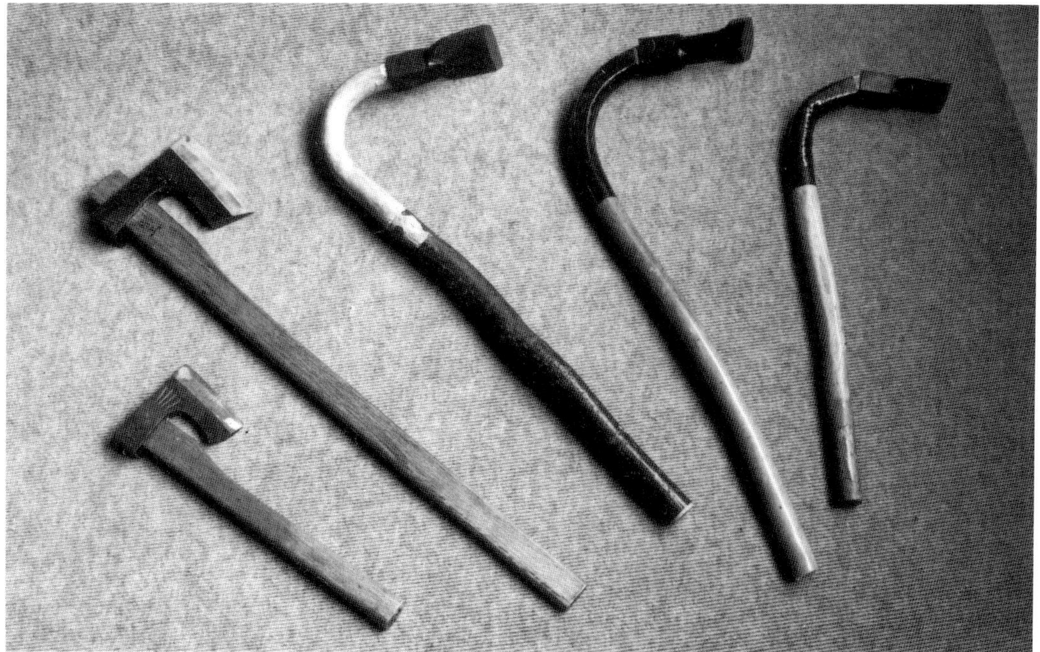

42. Axes and adzes. From left: two small axes and three adzes.

Saws

From the hefty ones wielded by lumberjacks to the tiny ones employed for detailed work in such tight places as the inside of wooden plane blocks, saws come in numerous sizes (fig. 43). The two main types, crosscut and ripping, are distinguished by the shape of their teeth (fig. 44). All cut on the pull stroke, which allows the blades to be relatively thin (since steel is strongest in tension) and places a smaller physical burden on the carpenter. As with all good Japanese tools, the best blades (and many of medium quality) are still made by hand; wooden handles and other fittings can be purchased, though they are preferably fashioned by the user to suit his own preference. Observes Nishioka, "Maybe Westerners think the added accuracy [provided by the thin blade of a Japanese saw] doesn't make any difference. . . . But then, an electric saw can cut pretty well, too, you know. It doesn't care if the wood is soft or hard!"

43. Saws. Top: crosscut saw for rough work. Bottom, from left: ripsaw for rough work; two single-edged crosscut saws; two double-edged saws for both ripping and crosscutting; extra-fine crosscut saw for cutting tenons; curved, double-edged saw for starting at the middle of a board; two "keyhole" saws.

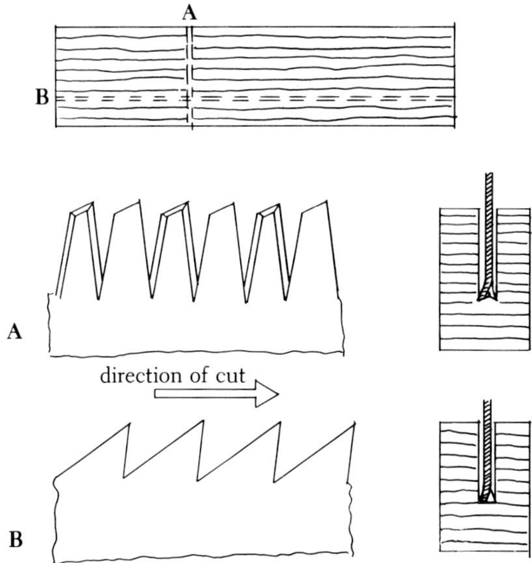

44. Saws. a) Crosscut: teeth have three distinct cutting surfaces (slightly different for hardwood and softwood). b) Ripping: similar to Western ripsaw (tooth angle slightly different for hardwood and softwood).

Chisels

There are several varieties of chisels, distinguished primarily by the cross section and length of the blade and by whether the chisel is designed to be struck with a mallet or pushed by hand (fig. 45). Ranging in width from a few millimeters to about five centimeters, blades are attached to the wooden handle by a thick tang. Two different steels are laminated to form the blade, a harder one for the actual cutting edge and a softer, more flexible type for the back (fig. 46). The resultant blade holds a polished edge well and resists cracking under stress.

Planes

Like saws, Japanese planes (figs. 47–50) work on the pull stroke. They consist of a laminated steel blade fixed in a wooden body. The "common" planes are of many sizes and proportions as well as details of sole design, depending upon whether they are intended to remove wood quickly or to produce smooth surfaces. There are other varieties designed for rabbeting, forming curves, beveling corners, and producing shaped moldings, dadoes, and so on. The smallest is the size of the little finger. A well-tuned plane is capable of removing shavings as thin as ten microns and is probably the most difficult tool to master.

45. Chisels. Top row, from left: two paring chisels or slicks, 200-year-old long-handled chisel, four butt chisels, two short butt chisels, two thin chisels. Bottom row, from left: four mortise chisels, two round chisels or gouges, long-handled detail or dovetail chisel, crooked or trowel chisel.

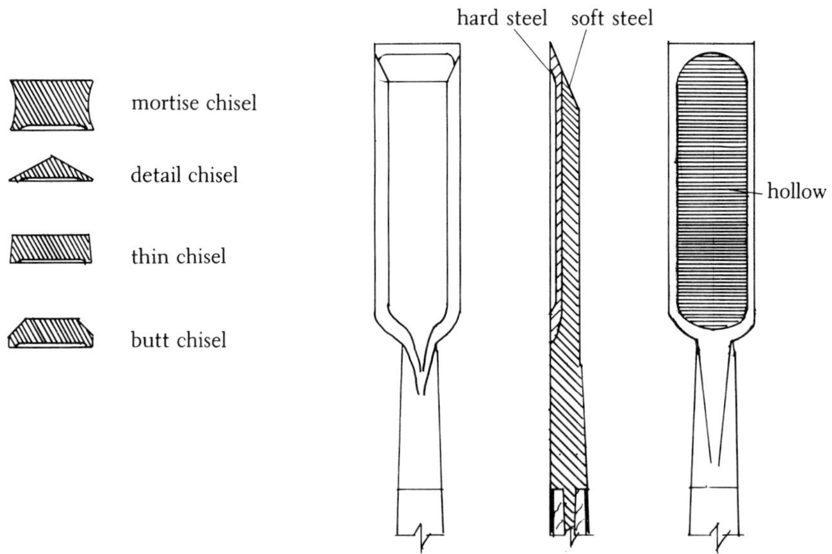

46. Chisels: many sizes and shapes for specialized uses. They are laminated of hard and soft steel.

47. Planes. Top row, from left: two long "common" planes, two short "common" planes, right- and left-hand corner planes. Bottom row, from left: miniature compass plane, miniature rounding plane, miniature blunt-nosed plane, scraping plane, plow plane, pointed side-shaving plane, rounding plane.

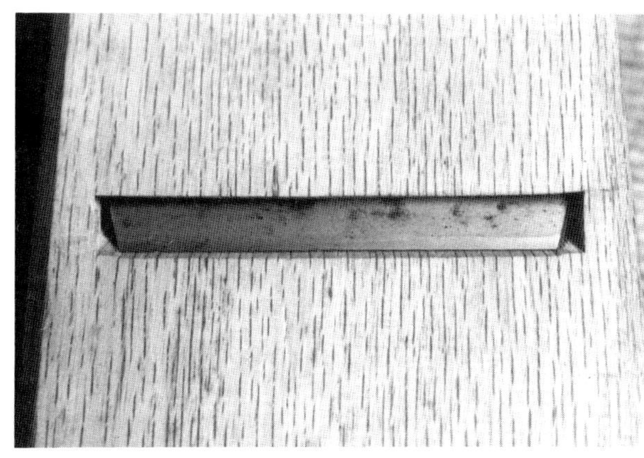

48. Detail of the sole of a plane, showing "mouth" and protruding blade.

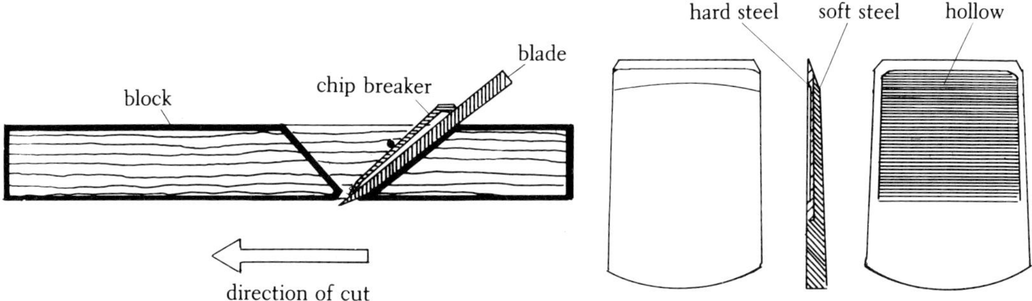

49–50. Planes. A laminated blade set in a wooden block.

Spear Planes

The "spear plane," or *yariganna*, is the ancestor of the present-day plane, which did not appear in Japan until the sixteenth century. Shaped somewhat like a spear (fig. 51), it is used in long, gentle, slicing strokes to smooth wooden surfaces, producing a slight undulation that is very soft to the touch. Nishioka revived the spear plane after centuries of disuse in order to recreate the original finish of the wood surface at Hōryūji and Yakushiji. It can now be found all over Japan. "The spear plane and the electric plane are designed to do the same job," Nishioka observes. "The electric plane removes wood more quickly, but the spear plane, even though it was designed over a thousand years ago, leaves a surface that stays smoother much longer. Because the cut is so clean across the cell walls of the wood, water doesn't enter and the wood resists mold" (Tsunekazu Nishioka, *Ki ni Manabe* [Tokyo: Shōgakkan, 1988], 30).

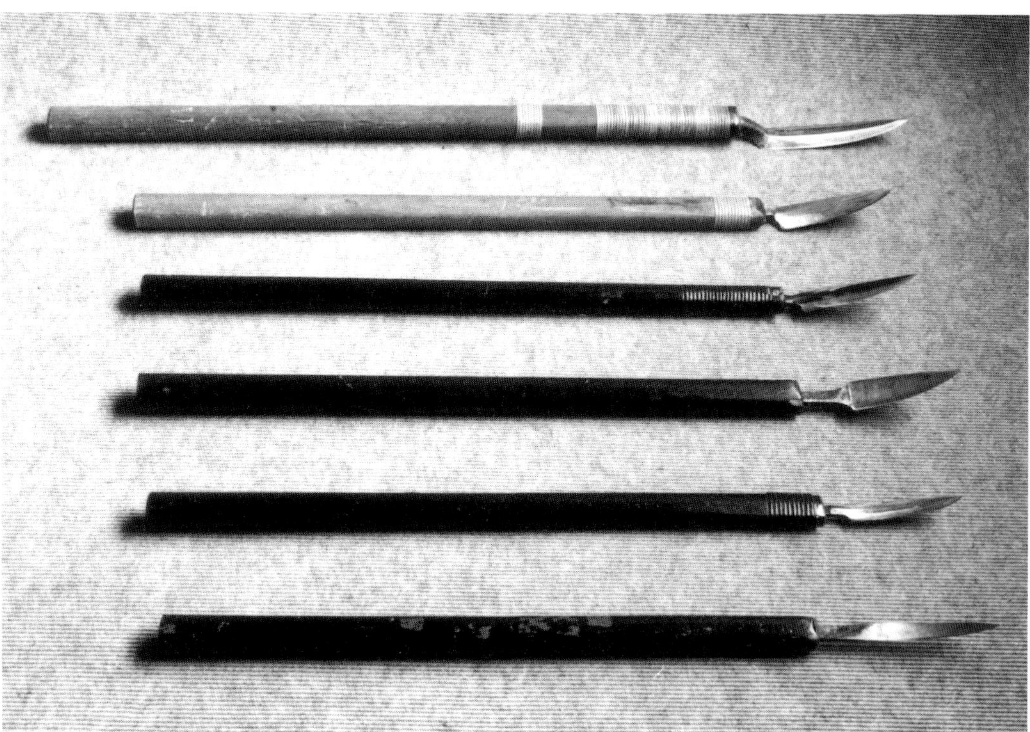

51. Spear planes.

Hammers

Hammers are employed primarily for driving chisels and secondarily for driving nails (fig. 52). Available in several weights and with a variety of head shapes (fig. 53), they exhibit a number of subtle design features and can be quite decorative. The largest are the sledgehammers for driving spikes and the massive mallets for nudging wooden members into place.

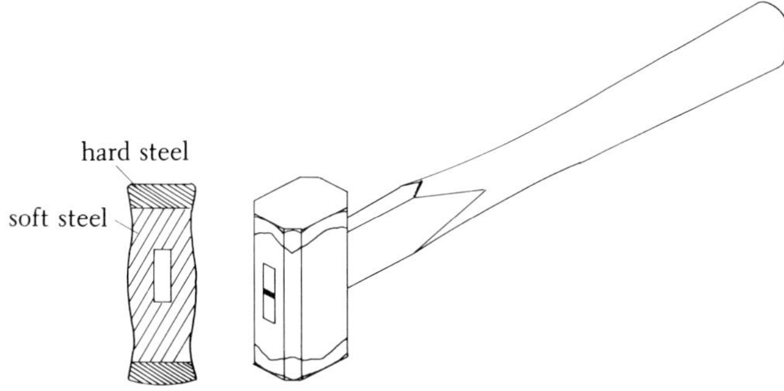

52. Hammer head. Flat face is used for striking chisels and driving nails; slightly curved face for sinking nails.

53. Hammers. Five types of hammers and a mallet.

Sharpening Stones

Sharpening stones (fig. 54), which are used with water, are found in varying degrees of coarseness. The stone is fitted to a wooden holder. It generally takes three stones of progressively fine grit to true an edge. Occasional resurfacing is done with a powdered abrasive on glass. At the end of the sharpening process, a steel plate is employed to achieve a high polish. In recent years, synthetic stones have become quite common.

54. Sharpening stones.

MAKING A JOINT

During the early periods of civilization in both Europe and Asia, the connection of wooden structural elements posed a great technological challenge. Throughout the world, stone-age people usually met this challenge by lashing timbers together with strong fibers such as vine or strips of tough bark. As stone tools developed, primitive joints were fashioned by gouging out holes for the insertion of roughly shaped tenons. With the advent of bronze tools, and later iron, wood joinery attained forms familiar even to us today, and experimentation led to the invention of splices, intersections, dovetails, and means of joining flat panels. In Western civilization, the Egyptians (as seen in their surviving wooden furniture) and the Greeks (as seen in their roof structures) are known to have possessed advanced wood joinery techniques; the Romans developed these further, building complex wooden geared water wheels, arched wooden centering for the construction of vaults and domes, and the familiar catapults and other machines of war. Most of these techniques were never entirely lost, surviving in Arab countries and reappearing in late Medieval Europe. Western wood joinery reached another peak in Tudor England, where it was often used in church spires and other monumental structures. Later English emigrants to America took with them a highly advanced repertoire of joinery techniques, most of which survived there, as in Europe, until the late nineteenth century. It is interesting to note that during this entire period the economic viability of wood joinery depended upon its low cost relative to other means of connection—metal fasteners such as bolts, nails, and the like—which had existed since the earliest eras but were often prohibitively expensive. In the West, this situation changed rapidly with the Industrial Revolution, when both cheaply milled lumber and mass-produced nails be-

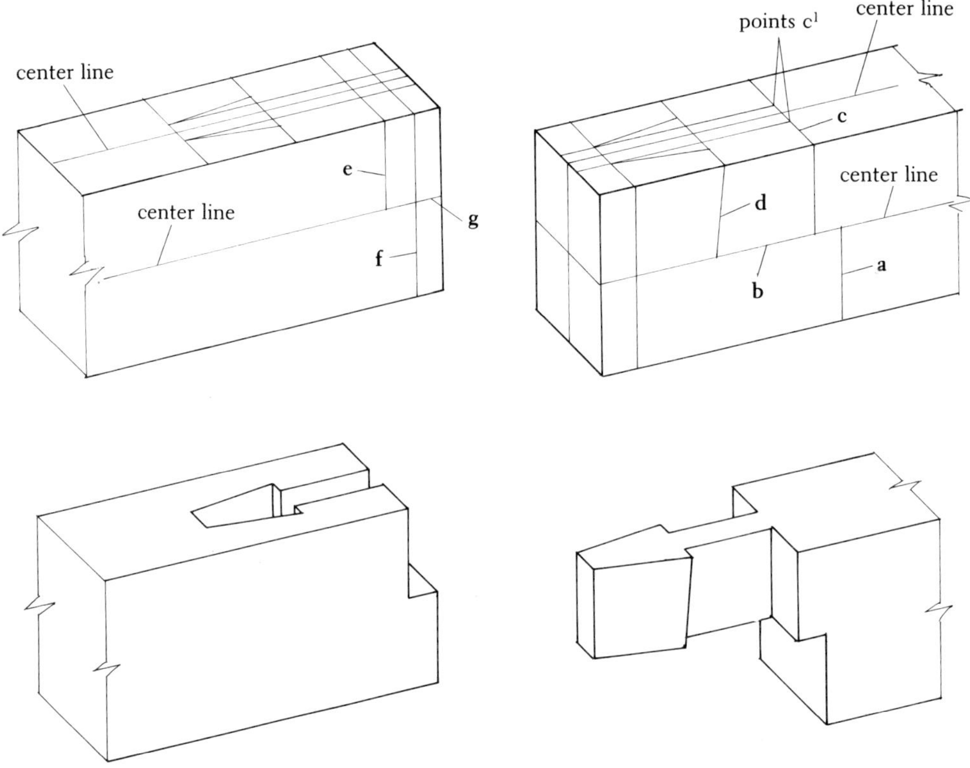

55. Half-lapped "gooseneck" splice joint. Top: joint laid out.
Bottom: completed mortise and tenon.

came common. By the early twentieth century, architectural wood joinery had virtually disappeared.

In Asia, the situation was somewhat different. It may be argued that even in the millennium before Christ, Chinese joinery technique was more advanced than its Western counterpart. Although dynasties changed frequently, there was a cultural continuity in all fields unmatched in European culture, reaching one peak during the Han dynasty (A.D. 0–200) and another during the T'ang (A.D. 600–900). The complex joinery practiced by Japanese carpenters during the building of early temples like Yakushiji represents the state of the art in early T'ang China, as distilled once by contemporary Koreans and again by builders on Japanese soil. Exactly how many of the early Japanese temple builders were native Japanese is open to debate. Nishioka has remarked that because Hōryūji temple is based on an ancient Korean *shaku* scale, and Yakushiji on a Chinese scale, their master carpenters may have been Korean and Chinese respectively.

TENON

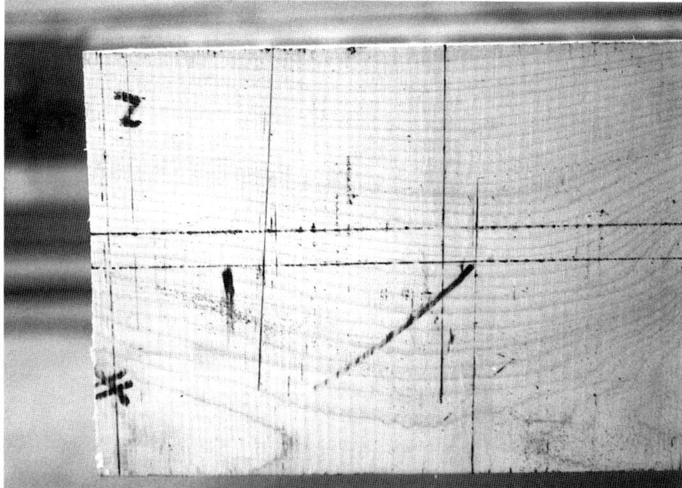

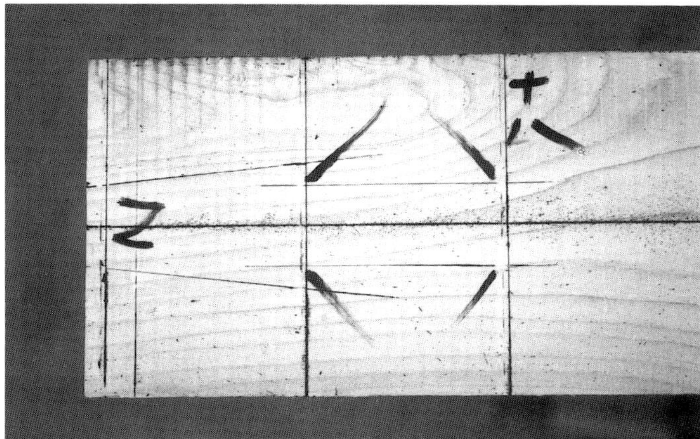

56–57. The joint is first laid out on the end of the timber. The most important line is the center line. The other dimensions are standardized and can be quickly laid out with a steel square.

The joinery seen in Japan today thus represents the survival of an extremely archaic way of building, a kind of living time capsule. As Nishioka says, the ancient carpenters were incomparably better, and yet, to this day, advanced joinery can be found throughout Japan. Most houses, when constructed in wood, feature traditional joints reinforced with hardware, thanks partly to newly developed computer-assisted milling machines that mass-produce jointed timber frames. Nevertheless, all agree that joinery's days are numbered. Prefabricated housing and two-by-four construction have proven to be economical, efficient alternatives.

Many of the techniques used in temple joinery can be demonstrated in the fabrication of a basic end joint or splice, in this case the half-lapped "gooseneck" joint, one of whose functions is to connect sections of a roof purlin (figs. 55–82).

58–59. The waste wood below the tenon is removed with a saw, first crosscutting (fig. 58) and then ripping (fig. 59). In both cases, the cut is taken right to line *a* and *b* (see fig. 55).

60. The waste is knocked out with a hammer, and the inner corner cleaned up with a chisel.

61. The underside of the tenon is laid out on the freshly cut surface with a steel square and bamboo marking pen.

62. The angled surfaces of the tenon are sawed, extending the line all the way to cheek line *c*.

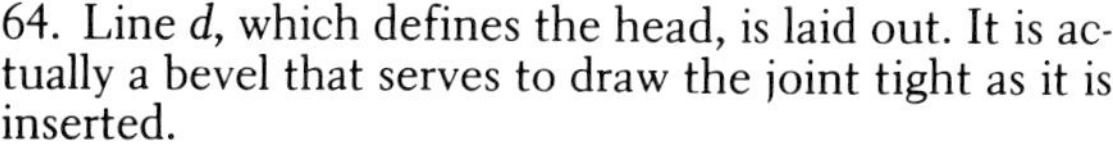

63. The cheeks are sawed all the way to the points marked c^{1}, *and the waste removed.*

64. Line *d*, which defines the head, is laid out. It is actually a bevel that serves to draw the joint tight as it is inserted.

65. The tip is trimmed to the line, and the bevel line *d* is cut. This is repeated on both sides.

66. The waste is chopped out with a chisel.

67. The surface is smoothed with a slick.

68. The finished tenon.

MORTISE

69. The mortise laid out (see also fig. 55).

70. The mortise is first drilled to remove as much waste as possible.

71. The lap at the end is cut with a saw, first crosscutting at lines *e* and *f* and then ripping along line *g*.

72. The waste is knocked out, and the inner edge trimmed with a chisel.

73. The mortise is laid out on the freshly cut vertical face.

74. The mortise sides are cut with a ripsaw as far as possible.

75. The mortise is chopped out with a chisel at an angle.

76. The mortise is chopped vertically.

77. At this point, a parallel-sided mortise is obtained.

78. The beveled line for the head of the tenon is laid out inside the mortise and then sawed.

79. The waste is chopped out with the chisel and ....

80. ... then the line is trued.

81. Lastly, the inner faces are smoothed with a slick, and the bottom surface given a slight incline to ease insertion; the surrounding edge of the opening on the upper face is also beveled slightly.

82. The finished mortise.

ERECTION OF THE PICTURE HALL

The following section deals with the demystifying experience of observing the actual assembly and erection of the new Picture Hall in the Sanzō-in compound of Yakushiji (figs. 83–88; see also figs. 17–18 and color pl. 2). On November 1, 1985, approximately six months after the fabrication of parts had begun, work was started on the erection of the frame. This stage of the work—installation of the columns, tie beams, and brace beams—took four days. The bulk of the roof structure was completed sometime before December 8, when the ridge-beam ceremony was held.

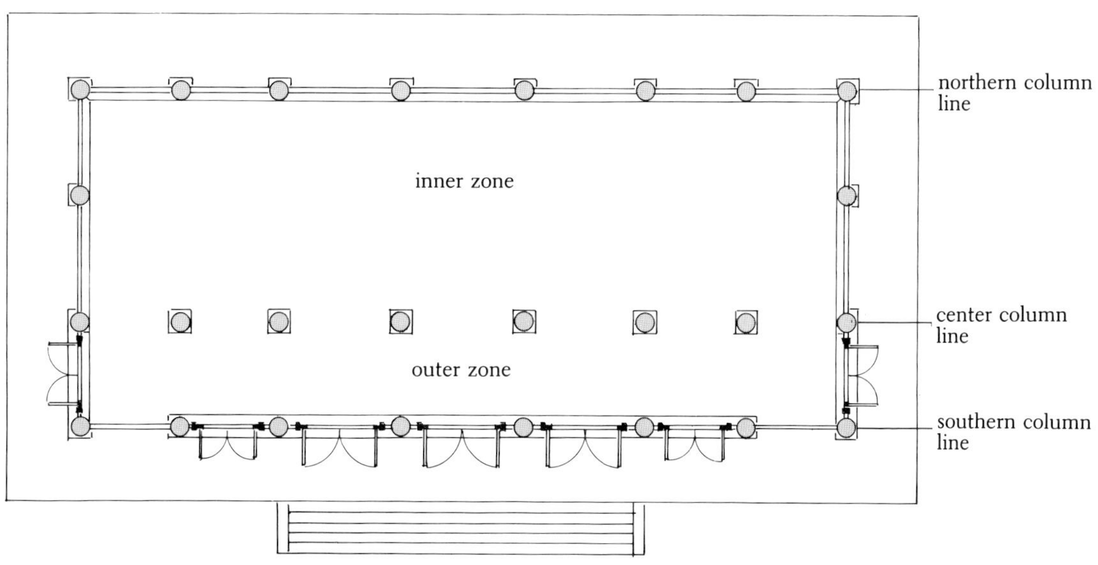

83. Plan of the Sanzō-in Picture Hall.

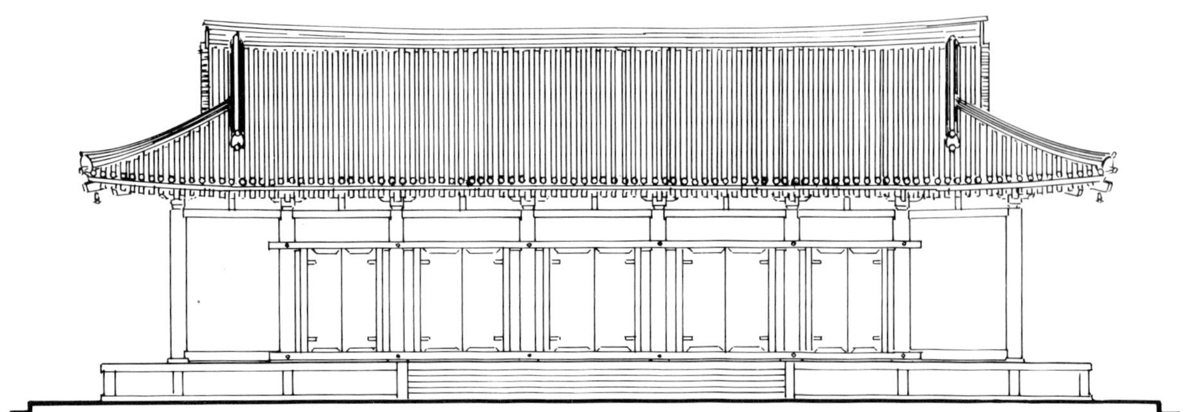

84. Southern elevation.

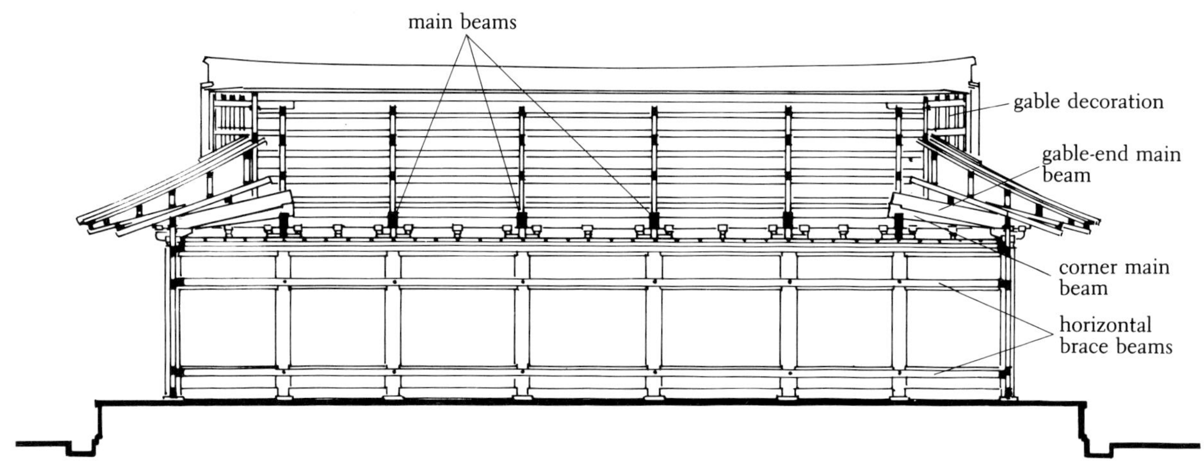

85. East-west section.

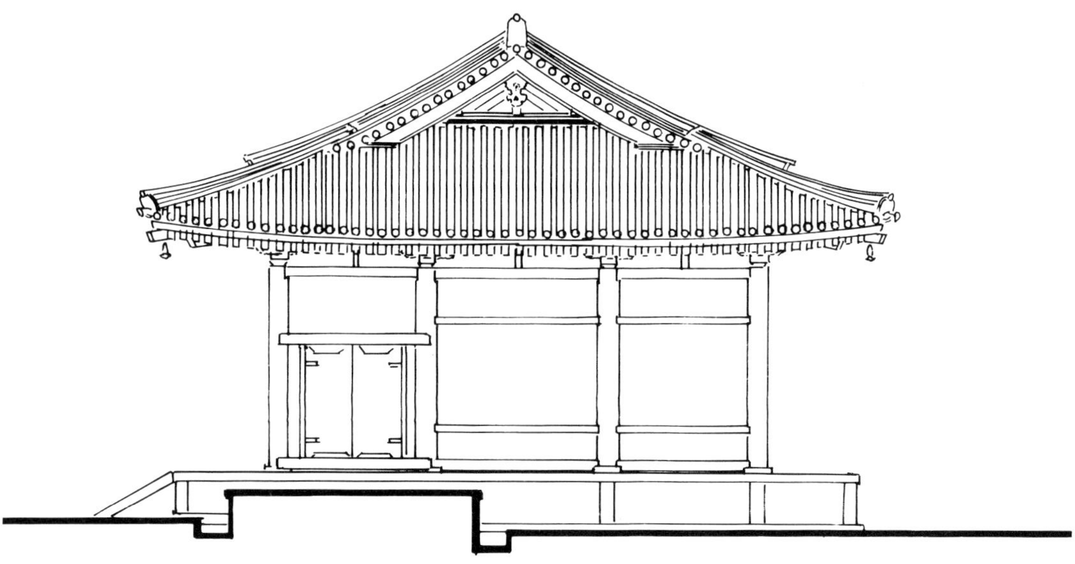

86. Eastern elevation.

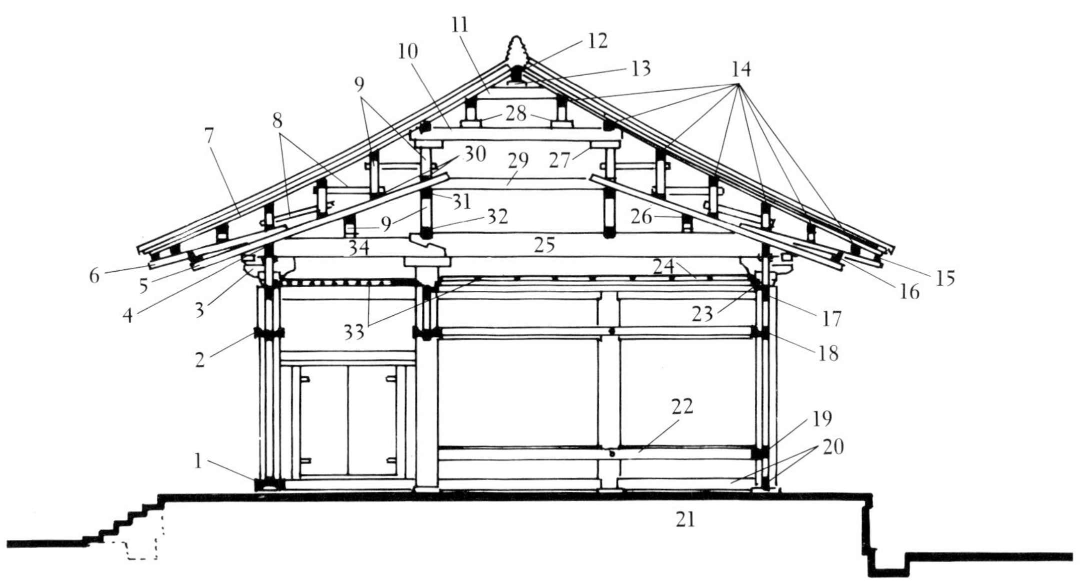

87. North-south section. 1) doorsill, 2) mid-wall horizontal brace beam, 3) bracket complex, 4) wall purlin, 5) lower rafter, 6) upper rafter, 7) hidden rafters, 8) bridging, 9) stub posts, 10) mid-crosstie, 11) upper crosstie, 12) hidden ridge beam, 13) hidden ridge-beam support block, 14) roof purlins, 15) eaves support, 16) upper rafter support, 17) head ties, 18) mid-wall penetrating tie beam, 19) waist tie beam, 20) base tie beam, 21) platform, 22) waist horizontal brace beam, 23) ceiling horizontal brace beam, 24) ceiling, 25) main beam, 26) outside longitudinal tie, 27) hidden support block, 28) post support block, 29) lower crosstie, 30) rafter hold-down beams, 31) upper longitudinal tie, 32) lower longitudinal tie, 33) ceiling lattice, 34) short main beam.

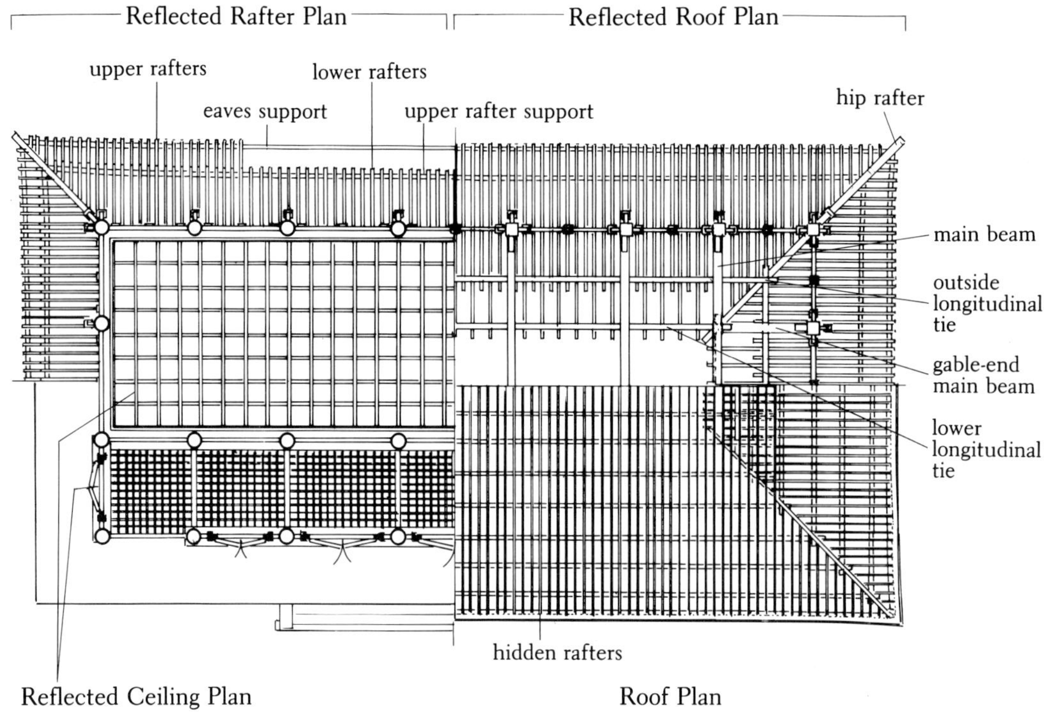

88. Reflected ceiling, rafter, and roof plans of Sanzō-in Picture Hall.

Thus the assembly of the basic structure, including most of the roof, required only one month. Thereafter, the size of the crew was reduced and work proceeded more slowly, with the last of the roofing tiles laid in May of 1986. Meanwhile, work had begun on the next building in the Sanzō-in compound, the octagonal hall, with only two carpenters remaining at the Picture Hall to work on the innumerable finishing details, such as the ceiling and doors. Everything was finally completed in January 1987 with the exception of the tile floor and interior paintings, which are expected to take until 1990.

During the construction of a building such as the Picture Hall, work proceeds simultaneously on several parts. This complicates the task of documenting the various stages. The main beams may already be in place at one end of the building, while the bracket complexes are not yet installed at the other. Meanwhile, work goes ahead on the horizontal brace beams, the doorsills, and even certain sections of the wall lattice. The following step-by-step explanation is given in logical sequence as far as the construction goes, but it does not necessarily adhere to strict chronological order.

Foundation, Platform, and Base Stones

The foundation is a heavily reinforced concrete platform set on footings 1.5 meters deep (figs. 89–91). Until modern times, foundations were simple stone-walled earthen podia into which were set large stones, one under each column. Today, largely for reasons of safety, modern foundation techniques are judged superior; fortunately the difference is undetectable to the eye.

When the building is complete, the rough foundation will be completely faced and paved in granite quarried in Hyōgo Prefecture (see figs. 181–83), but at the outset only the foundation stones are set. There are twenty-six foundation stones in all, each with a protruding bronze pin that will fit into a corresponding hole in the base of the column (see fig. 95). The bases of the columns themselves are slightly concave (fig. 92) to lessen the possibility of water being trapped beneath them and causing rot.

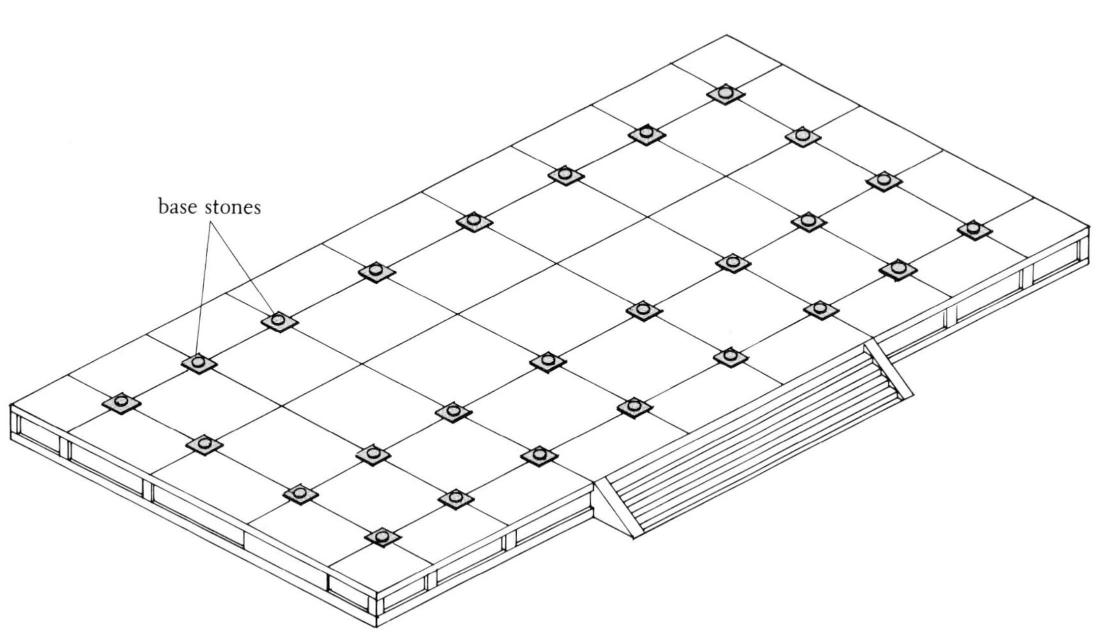

89. The Picture Hall after installation of foundation platform and column base stones.

90. Overall view of site with column base stones in place.

91. Masons position a base stone.

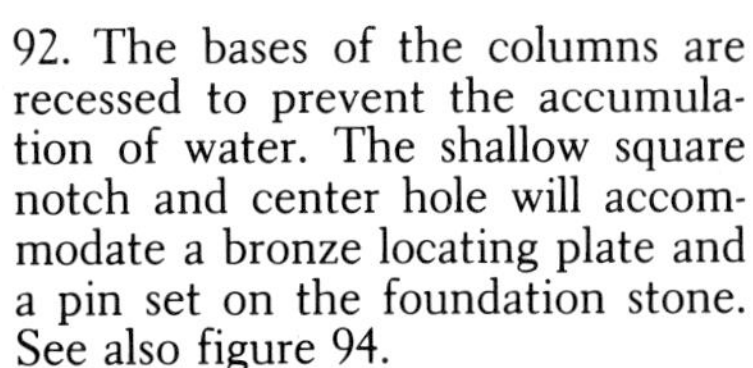

92. The bases of the columns are recessed to prevent the accumulation of water. The shallow square notch and center hole will accommodate a bronze locating plate and a pin set on the foundation stone. See also figure 94.

Columns and Penetrating Tie Beams

As illustrated in the plan (fig. 93), twenty columns form the perimeter, and six stand inside. All are tied together by penetrating tie beams at four levels: at the base, at waist height, at mid-wall height, and at the head of the columns. The inner columns do not have base ties or waist-height ties. For the perimeter columns, most of the base ties are trapped in notches at the extreme base end; those spanning door bays are inserted a few inches above the floor, leaving a small space. Since the base ties rest directly on the foundation (fig. 95), they are laid out first and the columns then positioned above, temporarily supported by wooden blocks (all with the help of a small crane; fig. 96). Mid-wall tie beams and door frames are inserted where necessary and secured lightly with wedges. Finally, the head tie beams are placed into slots at the uppermost ends of the columns. When a group of eight columns is assembled, the blocks are gradually removed and the entire unit is slowly lowered onto the foundation. The tie beams are more firmly wedged in, and the remaining columns and tie beams are added until the entire lower frame is complete. This process, from the erection of the first column to the completion of the lower frame, took four days to complete.

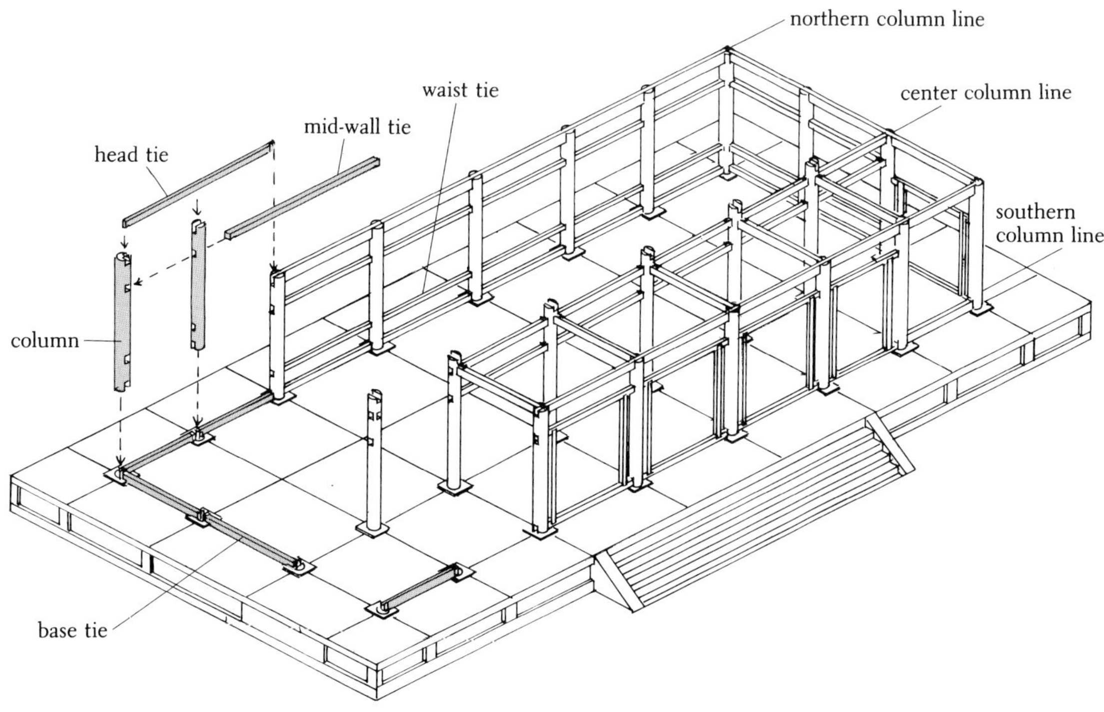

93. The Picture Hall with lower frame nearly completed.

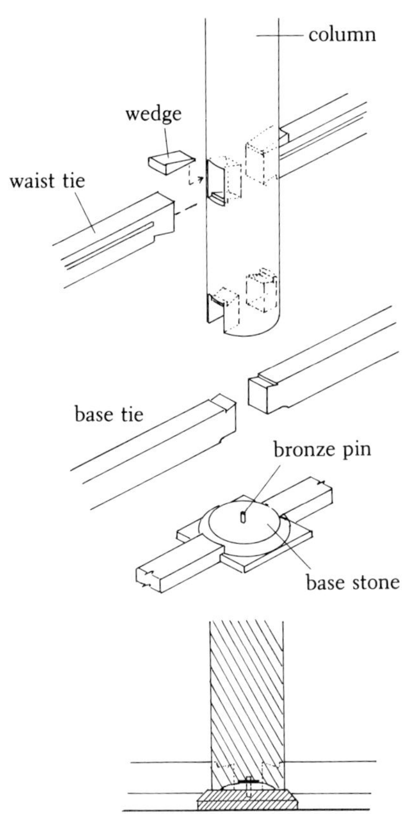

94. Configuration at base end of column.

95. Base tie beams resting on the base stone. Bronze locating pin for column is visible at center.

96. Mid-wall penetrating tie beams are inserted into the columns on the ground, and then raised as a section.

The uppermost tie beams, or head ties, are connected by splice joints located over the columns, taking advantage of the reinforcement provided by the surrounding slot (figs. 97–98). At the corners, they penetrate only partway into the column. The outer zone between the southern line of columns and the center line columns are also spanned by head ties, which require a 90° intersection.

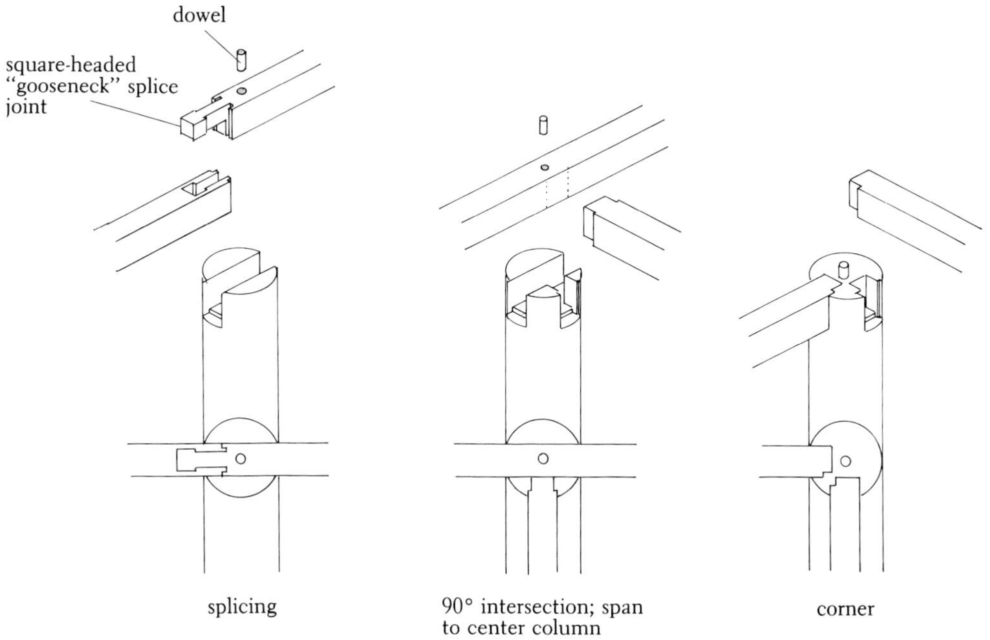

97. Relationship between head tie beams and columns.

98. Square-headed "gooseneck" splice used on head tie beam where it joins over column.

Outside Horizontal Brace Beams

Decorative horizontal braces are added above and below door openings (on both the interior and the exterior) and at ceiling height, mid-wall tie height, and waist-tie height on the interior (fig. 99). These braces have semicircular notches which fit around the columns, and each segment must be fitted individually to the curves of the columns after the basic frame is erected. The brace beams are jointed and must fit without gaps.

A special kind of brace beam is used at the bottom of the doorways (figs. 100–101). As mentioned on page 68, the doorsill is designed for easy replacement.

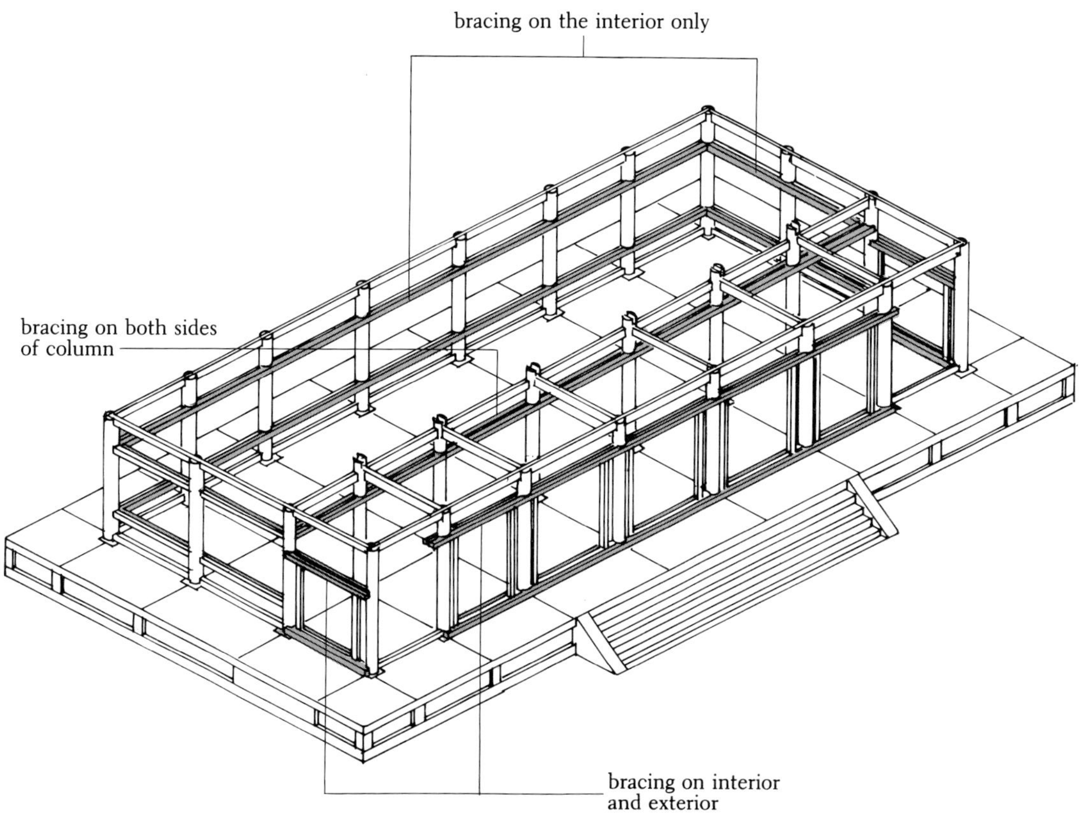

99. The Picture Hall after installment of bracing.

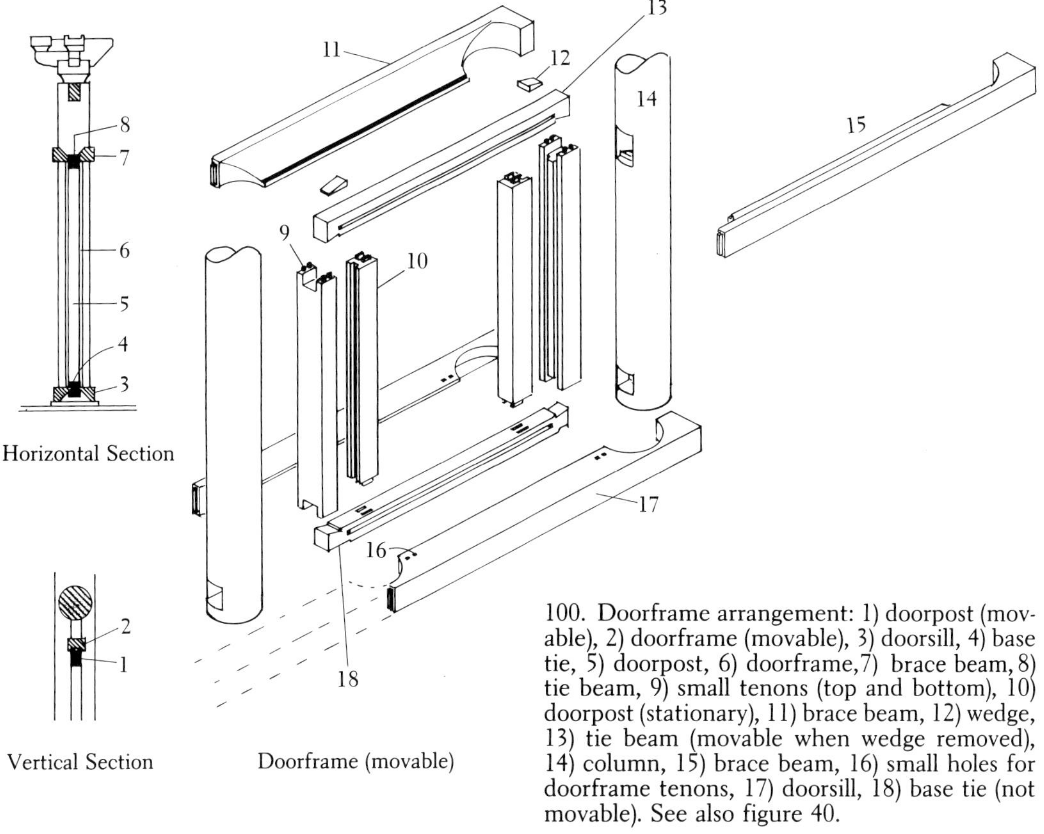

100. Doorframe arrangement: 1) doorpost (movable), 2) doorframe (movable), 3) doorsill, 4) base tie, 5) doorpost, 6) doorframe,7) brace beam, 8) tie beam, 9) small tenons (top and bottom), 10) doorpost (stationary), 11) brace beam, 12) wedge, 13) tie beam (movable when wedge removed), 14) column, 15) brace beam, 16) small holes for doorframe tenons, 17) doorsill, 18) base tie (not movable). See also figure 40.

101. Doorsill being installed.

Bracket Complexes

One bracket complex is located above each of the perimeter columns (fig. 102). Altogether there are fourteen regular, four corner, and two special bracket complexes (these last two will be discussed in the next section). For each regular bracket complex (figs. 103, 105), a large block is first fixed atop the column by means of a round dowel inserted into the head tie. Next a crossarm is added, and then a center bracket arm, which rests on both the large block and the crossarm. Finally, two small blocks are placed atop the crossarm, situated on round pegs, to complete the bracket (fig. 107, color pl. 4).

The corner bracket sets are similar (figs. 104, 106), except that a tenon replaces the dowel and two crossarms are placed on top of the large block. The center bracket arm is shaped to fit over the crossarms, ultimately nestling among four small blocks fixed by dowels to the two crossarms (fig. 110).

At this time, short block-bearing struts are put in place between each bracket complex (figs. 102, 111, color pl. 6).

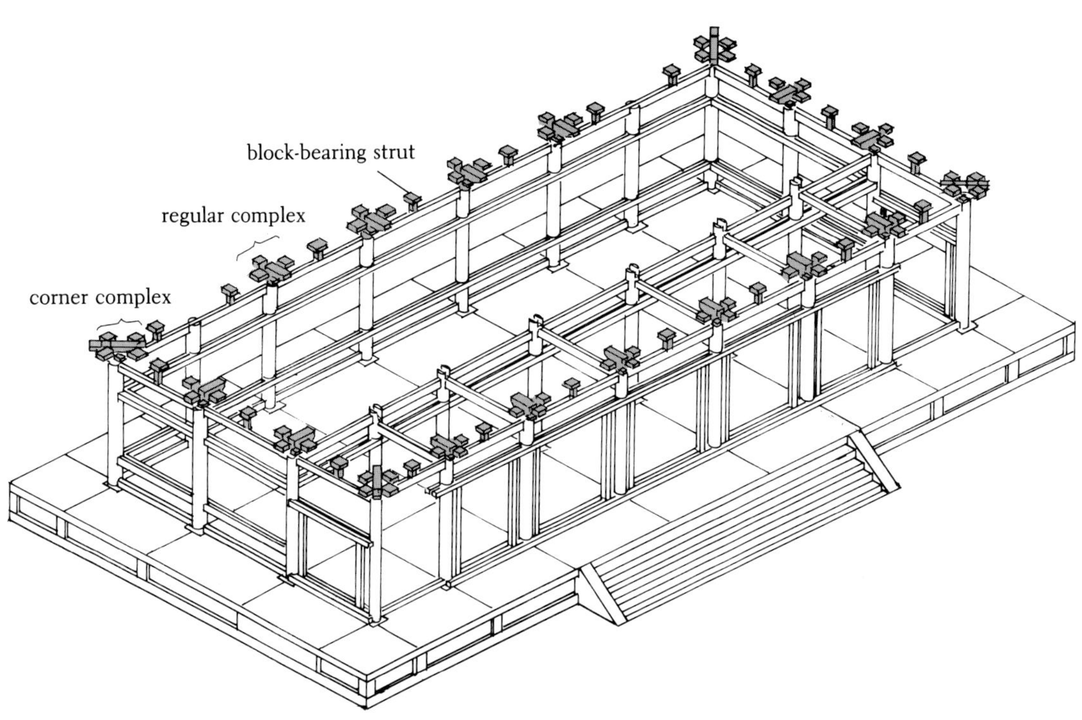

102. The Picture Hall after installation of bracket complexes to support roof load.

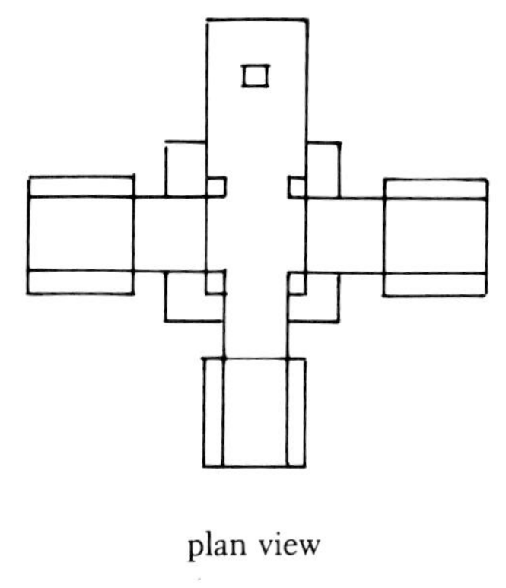

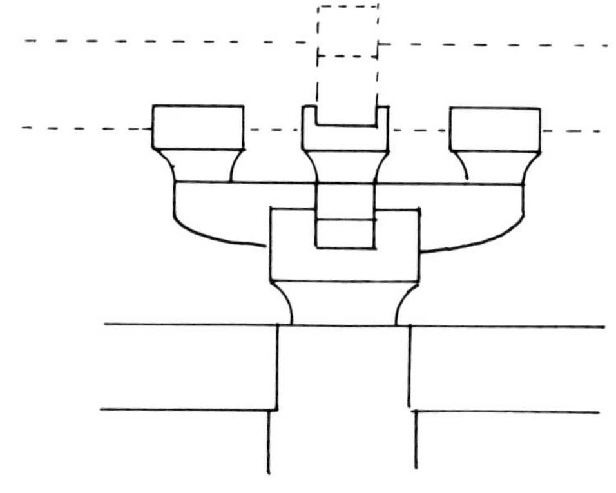

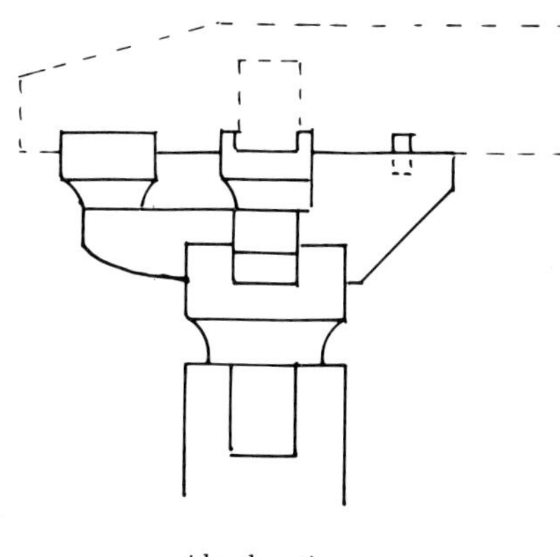

103. Regular bracket complex.

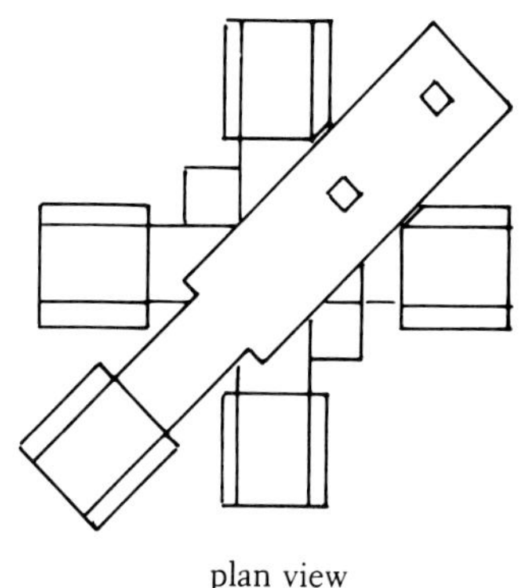

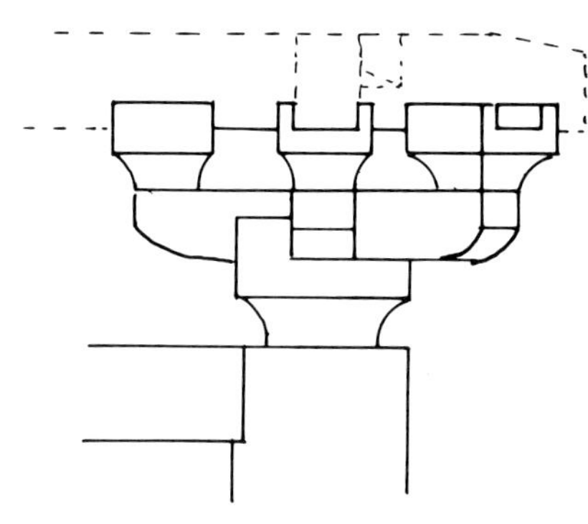

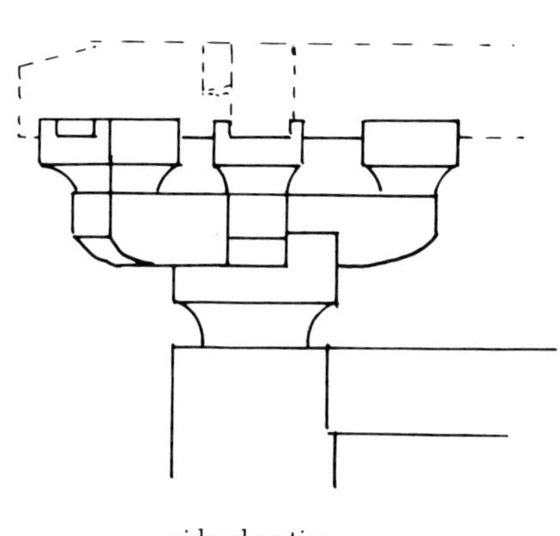

104. Corner bracket complex.

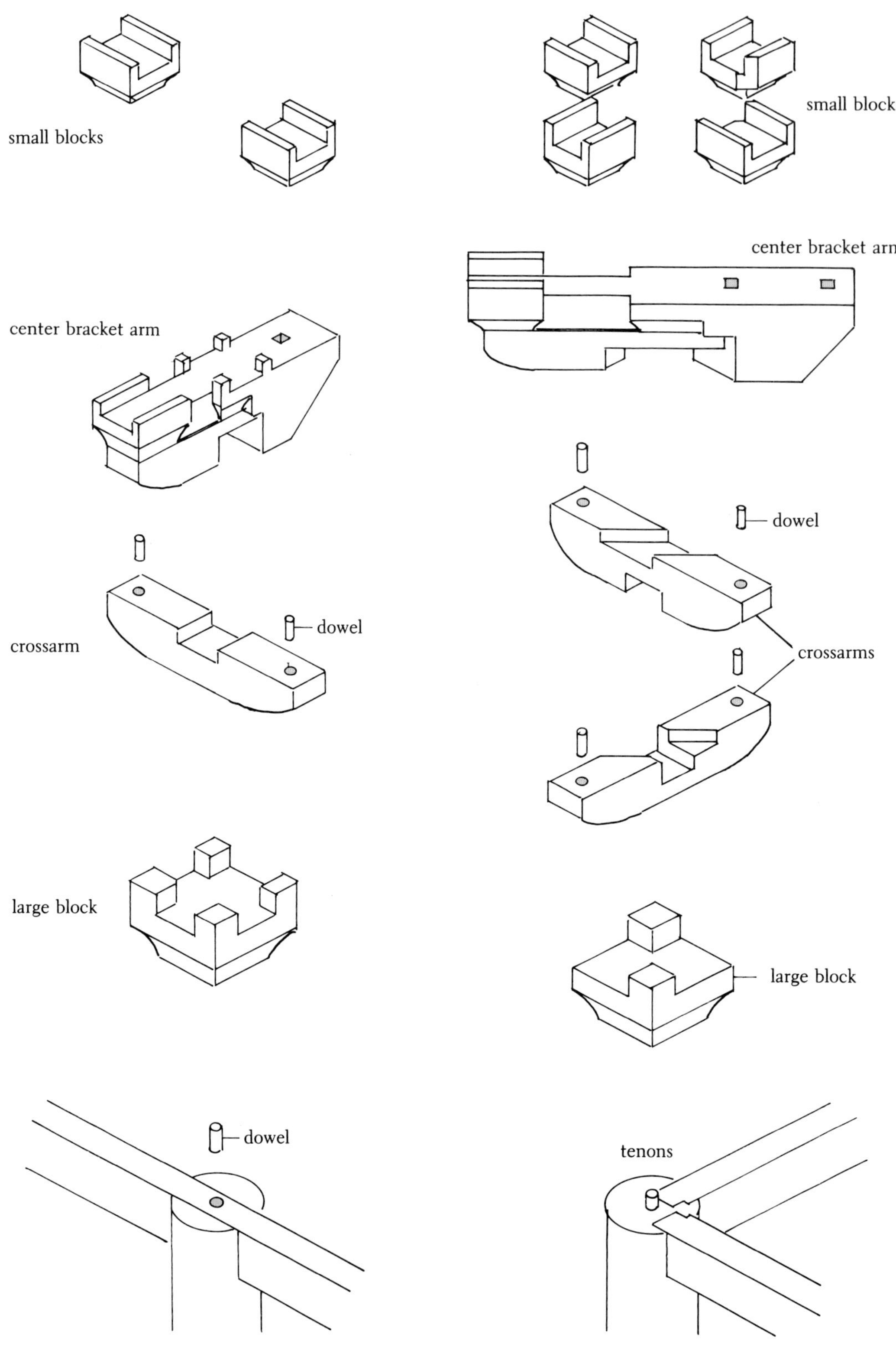

105. Regular bracket complex assembly.

106. Corner bracket complex assembly.

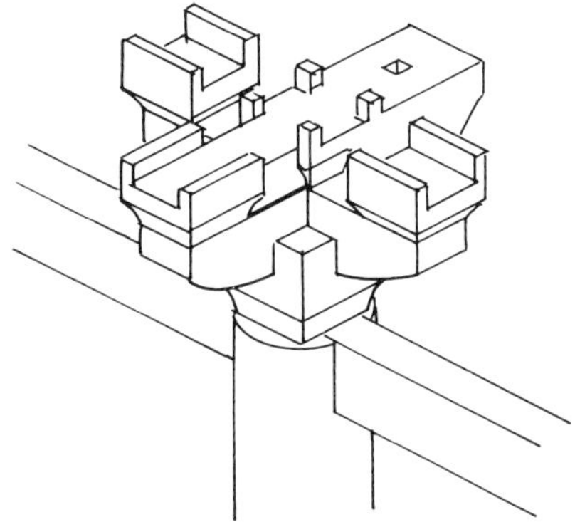

107. Completed regular bracket complex.

108. Large block is located by means of a dowel inserted into the head tie beam over center line of column.

109. Center crossarm of a regular bracket complex fits over smaller crossarm, both firmly supported by the large block.

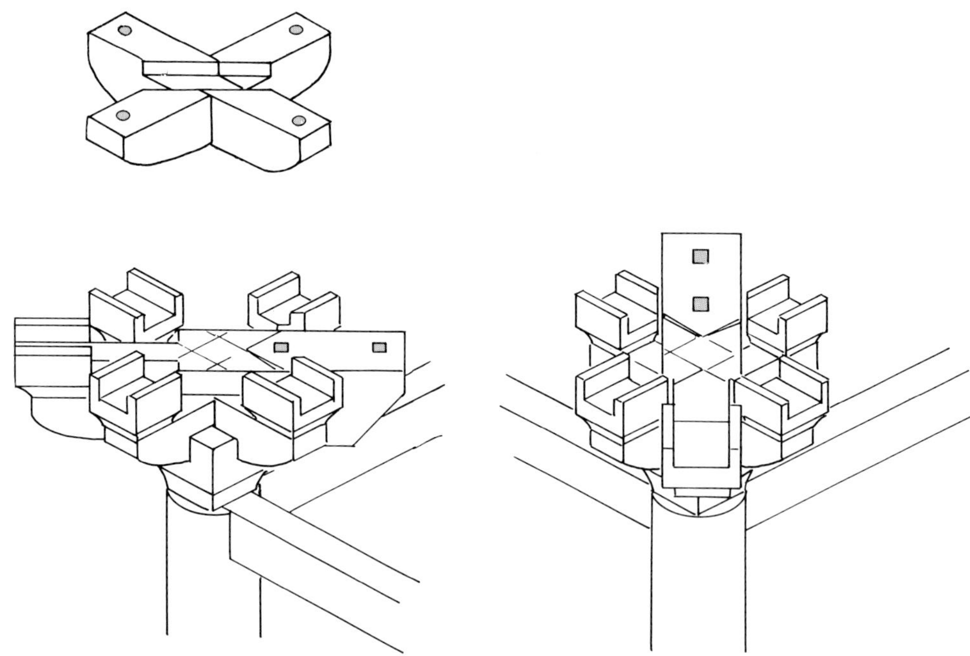

110. Completed corner bracket complex.

111. Completed corner bracket complex.

Bracket-Beams

For reinforcement at the gable ends, two long beams with a bracket complex at one end span to the center column line (fig. 112). At the bracket end (fig. 113), a projection penetrates a specially shaped large block and rests directly on top of the column. Together with the crossarm, this projection locks the beam in place.

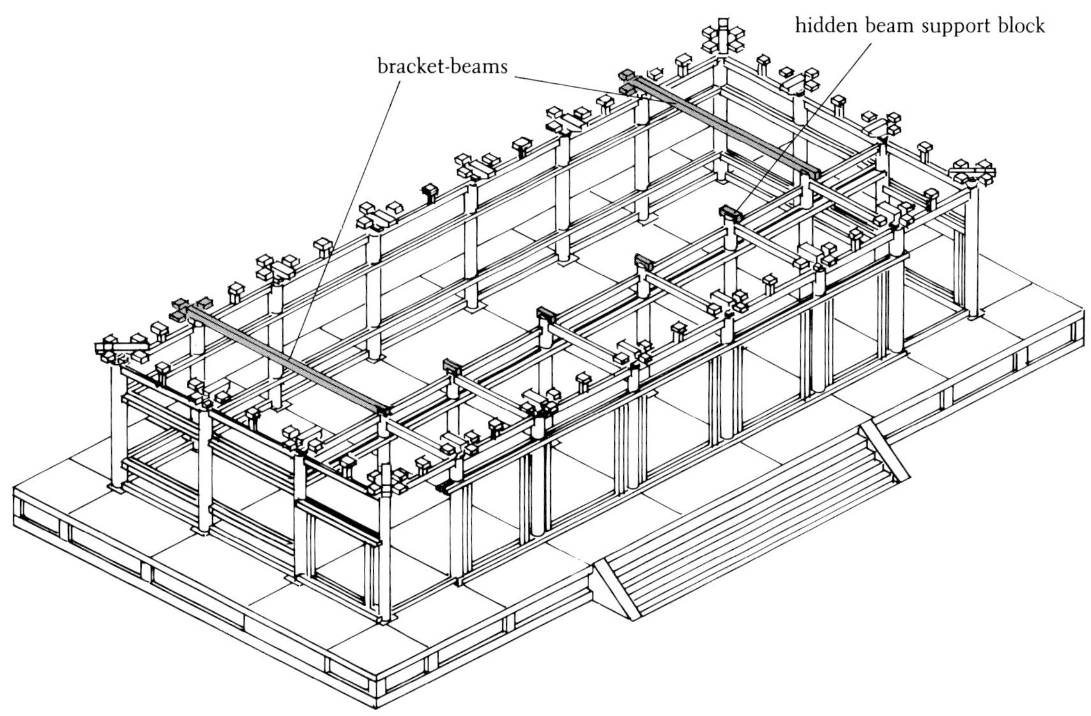

112. The Picture Hall after installment of bracket-beams.

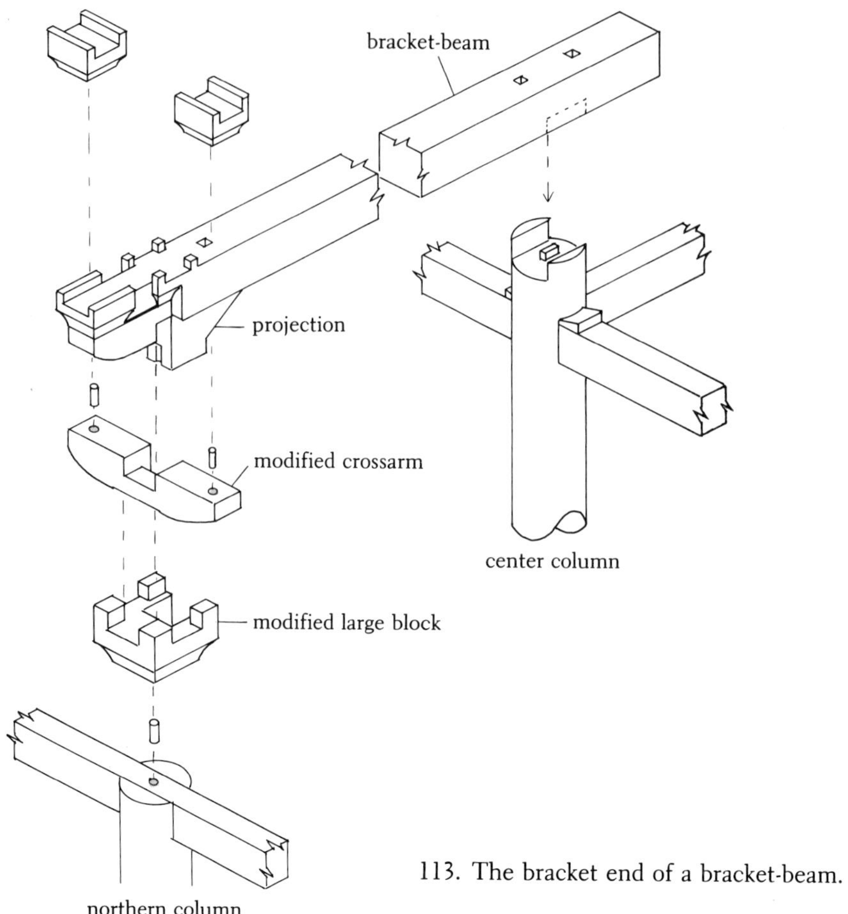

113. The bracket end of a bracket-beam.

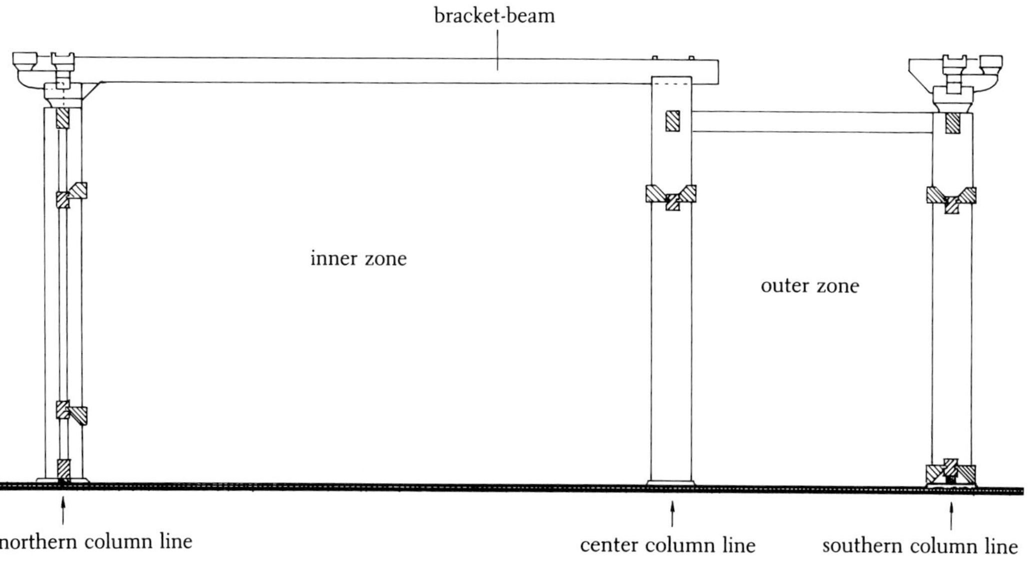

114. Sectional view of bracket-beam.

115. The bracket-beam has a projection which rests directly on the column and head tie beam, sharing the dowel with a modified large block.

116. The modified large block and crossarm are designed to be inserted after the bracket-beam has been emplaced.

Wall Purlins

The wall purlins tie the entire structure together at this level, resting atop bracket complexes and block-bearing struts (fig. 117). Together with the penetrating ties and head ties, they define the wall plane (see fig. 87). The wall purlins are jointed into three sections on the long sides and two on the short, with the joints falling above the bracket complexes and the columns (figs. 118, 123 color pl. 5). The joints are designed to provide a smooth bearing surface for the members which will rest on them, and to resist both tension and torsion. The wall purlins overlap at the corners (figs. 119–20, 122). Here we can begin to see the roof curve develop; as mentioned on page 66, while the head tie beams are perfectly horizontal, the wall purlins slope upward at the ends, supported by differences in height of the bracket sets and blocks.

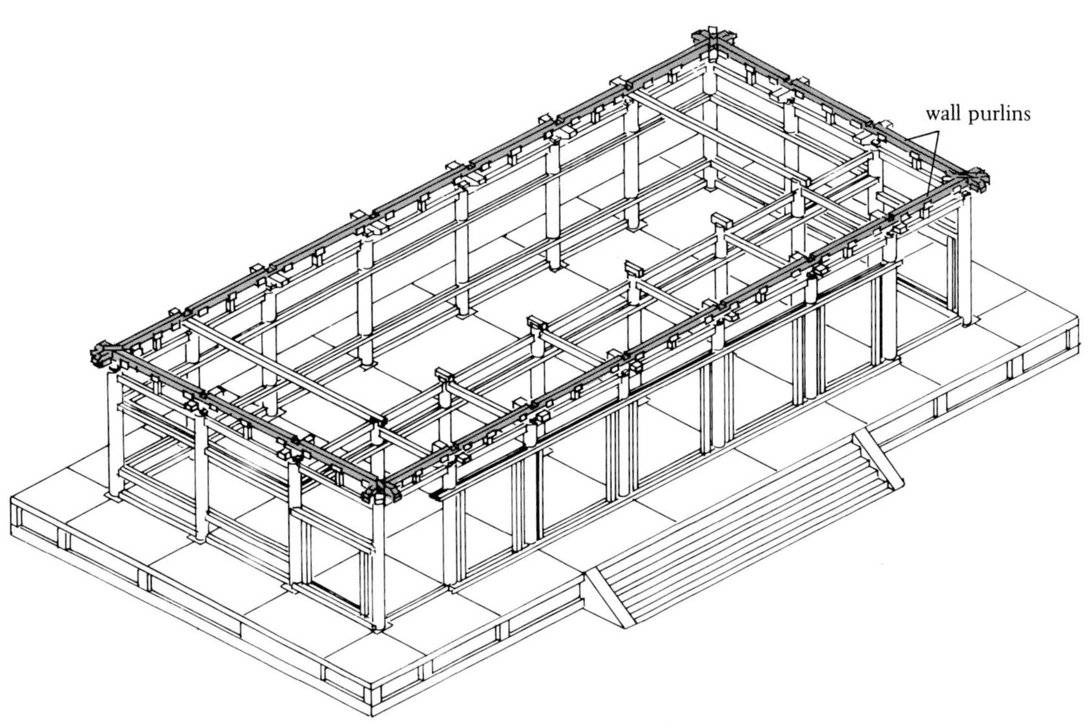

117. The Picture Hall after installation of wall purlins.

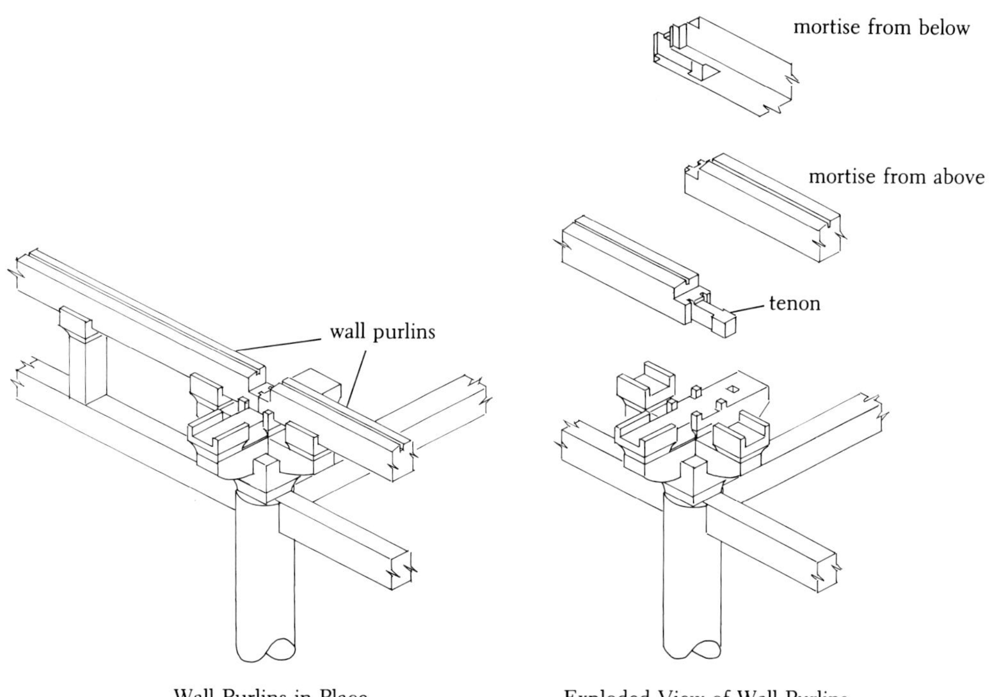

Wall Purlins in Place

Exploded View of Wall Purlins

118. Wall purlins are spliced over column and bracket sets with a variant of the "gooseneck" joint, leaving a recess and smooth surface on top to receive members above.

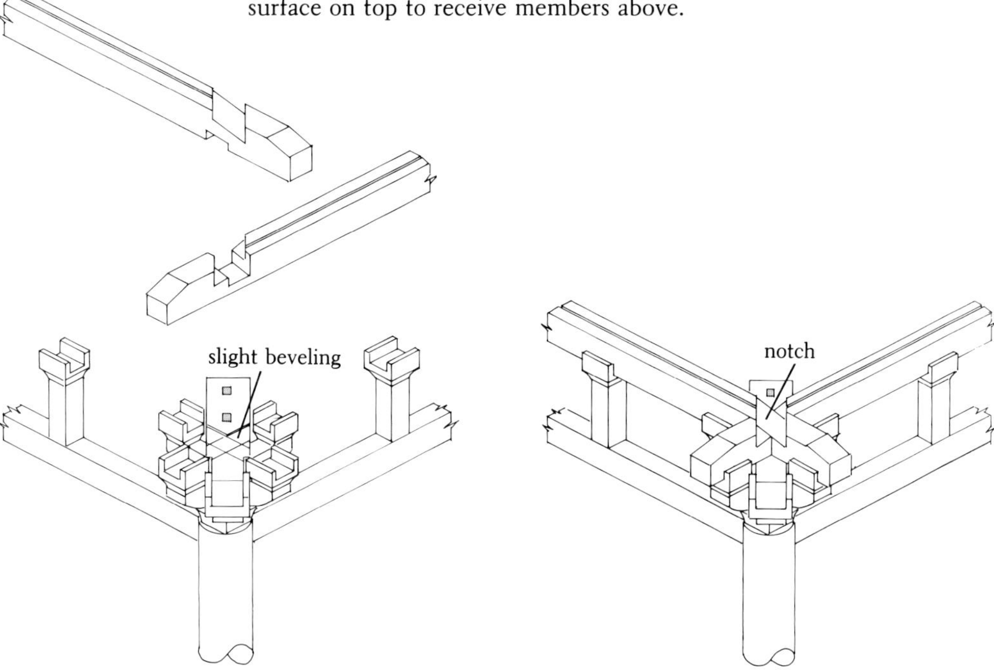

119. Wall purlins are joined at the corners with a simple cross-lap joint so as to accommodate the slope of the purlins as well as that of the members to be installed above.

120. Overlapping of wall purlins at corner. Note notch to receive upper hip rafter.

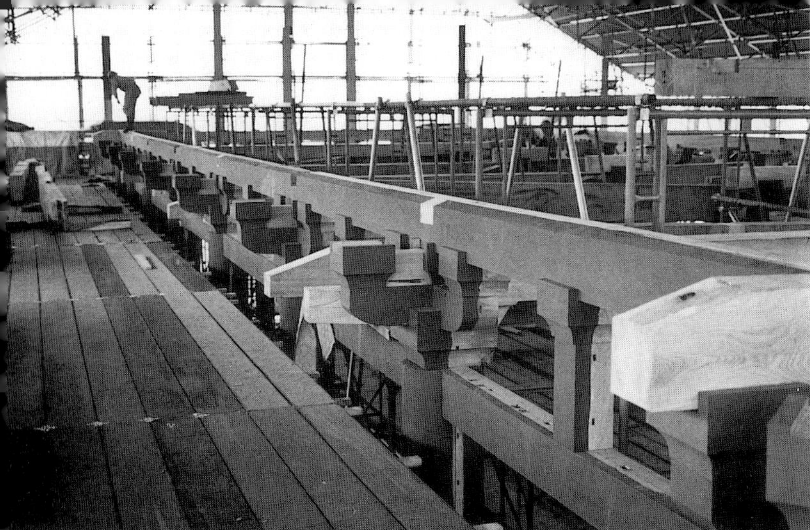

121. The slight upward sweep of the wall purlins at the ends is perceptible in this photograph.

122. Wall purlins overlapping at the corner.

123. The modified "gooseneck" splice of wall purlins over a bracket complex.

Main Beams

This next layer of structure is composed of heavy main beams which span the structure crosswise, bearing most of the roof load (fig. 124). These beams are in two sections of different length which join above the inner columns.

The shorter beams are installed first, with the help of a chain hoist. One end lies atop the block on an inner column, and the other end straddles the wall purlin situated as described earlier, resting on the bracket complex center arm (figs. 125–26).

The longer beams are similarly installed, joined to the shorter ones with a zigzag splice joint (fig. 127).

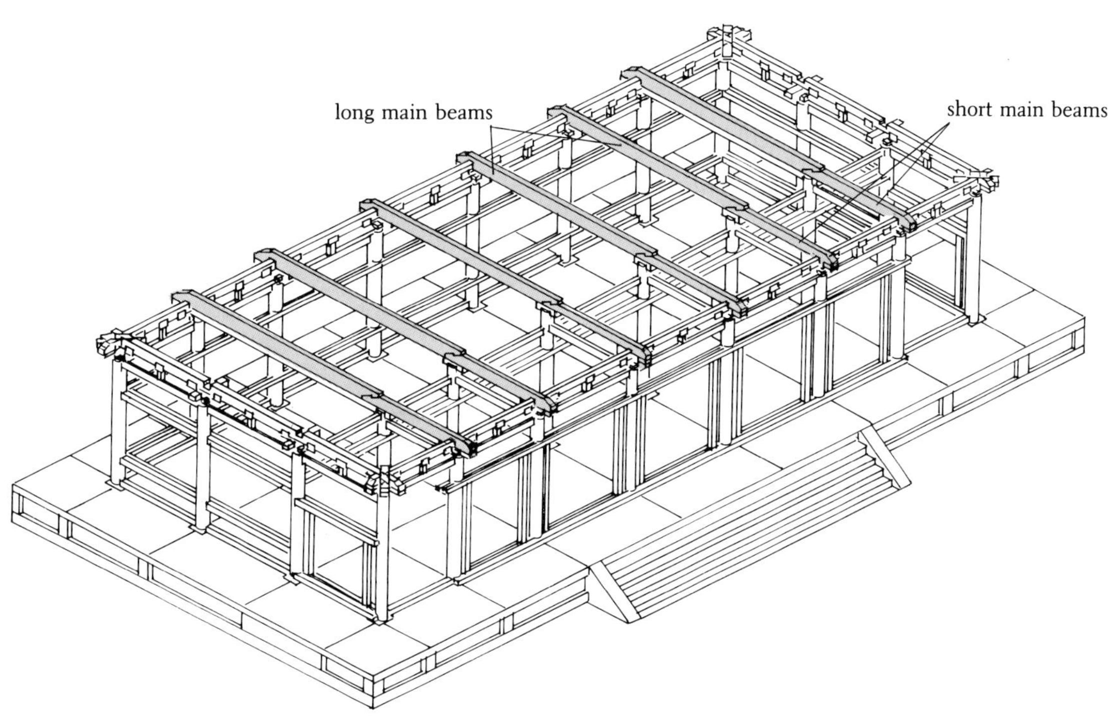

124. The Picture Hall after installation of the main beams.

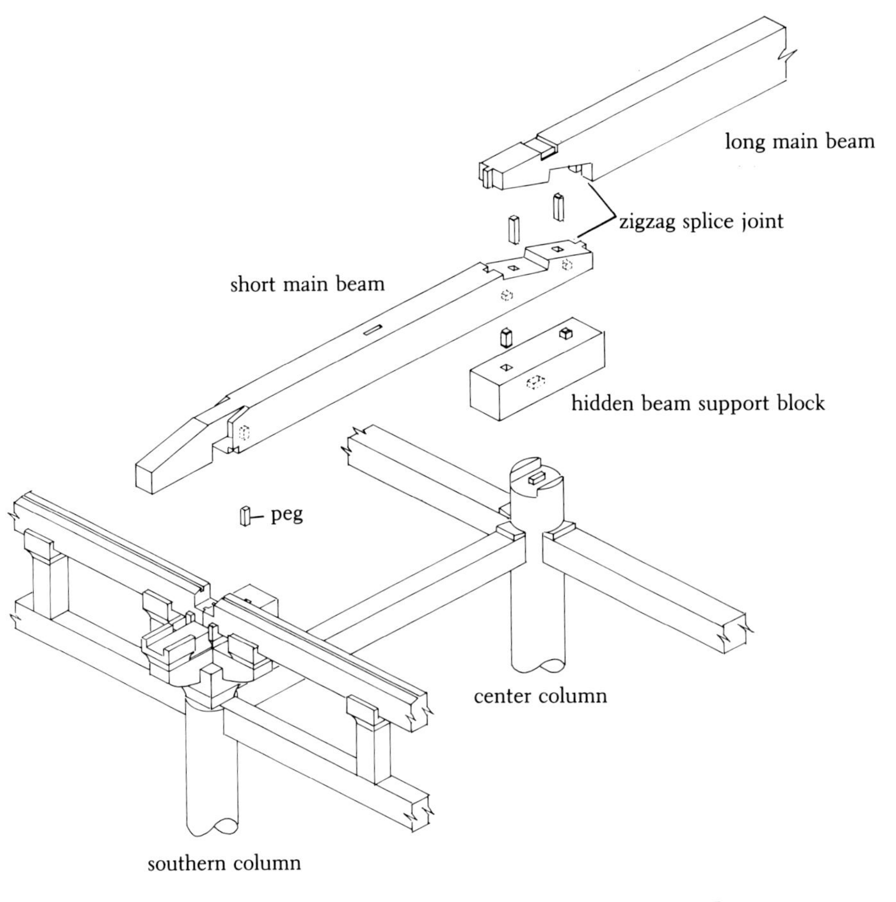

125. Main beam assembly.

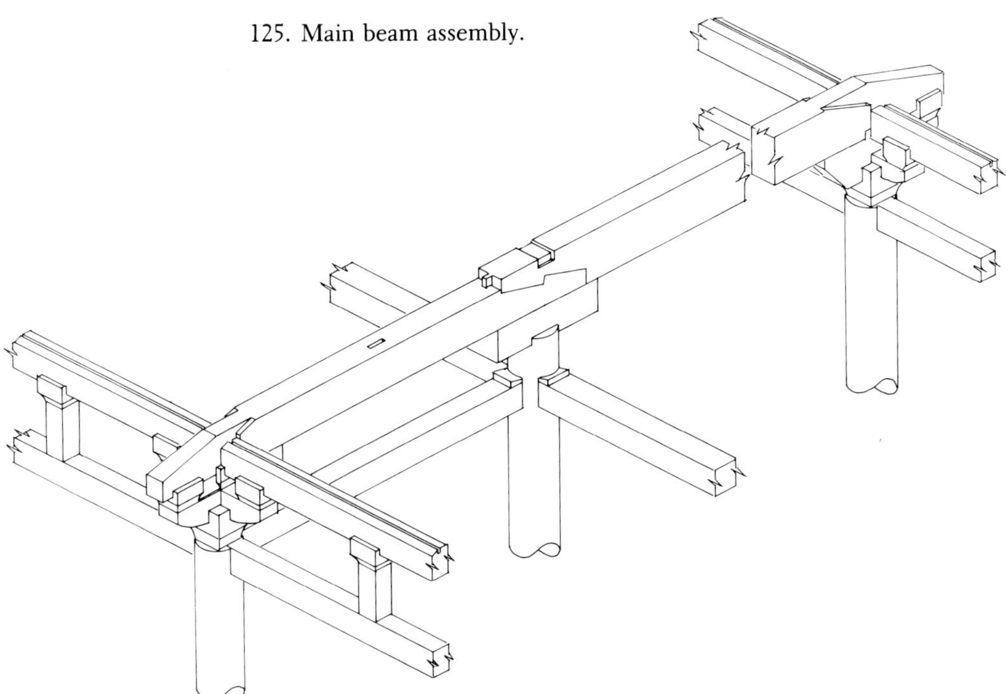

126. Completed assembly of main beam.

127. Zigzag splice joint.

128. The long main beams fit over the wall purlins and are supported by the center arms of the bracket complexes. Thus, they lock the entire structure together crosswise at this level.

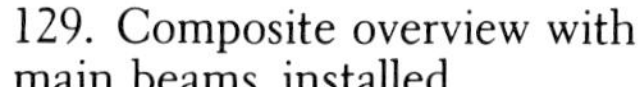

129. Composite overview with main beams installed.

Corner Main Beams, Gable-end Main Beams, and Lower Longitudinal Ties

Near the ends of the building, the long main beam rests directly on the long bracket beam, in effect becoming one extremely deep beam (fig. 130). This is necessary because the weight of the entire roof at the gable end will eventually come to bear here, transmitted through the corner main beams and the gable-end main beams.

Before the corner and gable-end beams can be installed, however, the other main beams must be tied together by means of long tie beams (lower longitudinal ties) which run almost the length of the structure. Like the head ties, these longitudinal ties are composed of jointed segments (see fig. 134) and sit in shallow notches on the upper sides of the main beams. They have mortises for short posts which will be added later.

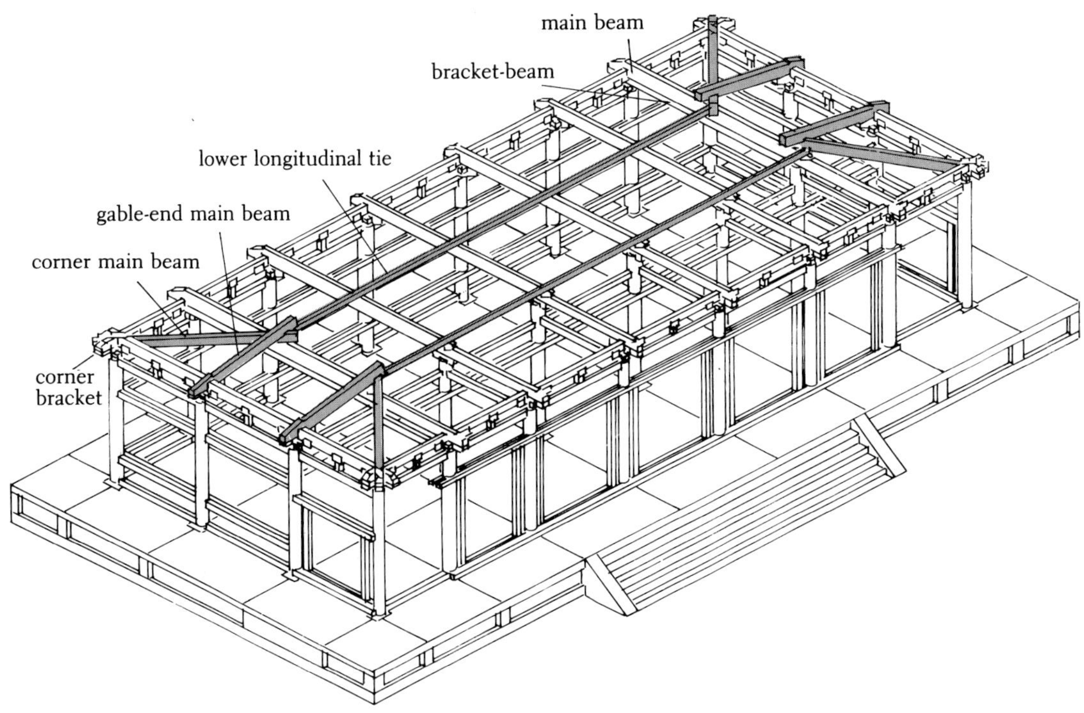

130. The Picture Hall after installation of corner main beams, gable-end main beams, and lower longitudinal ties.

The gable ends require a corner main beam which spans from the corner bracket set to the doubled-up main beam nearby, sloping downward (fig. 131). On top of it is placed the gable-end main beam, which slopes even more steeply, spanning to a bracket set (fig. 132, color pl. 7). The result is a pile-up of structural members above the main beams, creating a giant knot of timber.

131. The corner main beams span at a slope from the nearest main beam to the corner bracket complex. The gable-end main beams rest directly on the corner main beams and span to the exterior wall.

132. Like the other main beams, the gable-end main beam is installed over the wall purlin and bracket complex. The carpenter at left has one foot resting on part of the bracket complex.

Stub Posts, Outside and Upper Longitudinal Ties, and Crossties

Two more sets of longitudinal ties (the outside and upper) are installed here, supported on stub posts (fig. 133). They are then tied crosswise by crossties. These ties are also connected at the gable ends. The structure above an inner column at this point is as shown in figure 136.

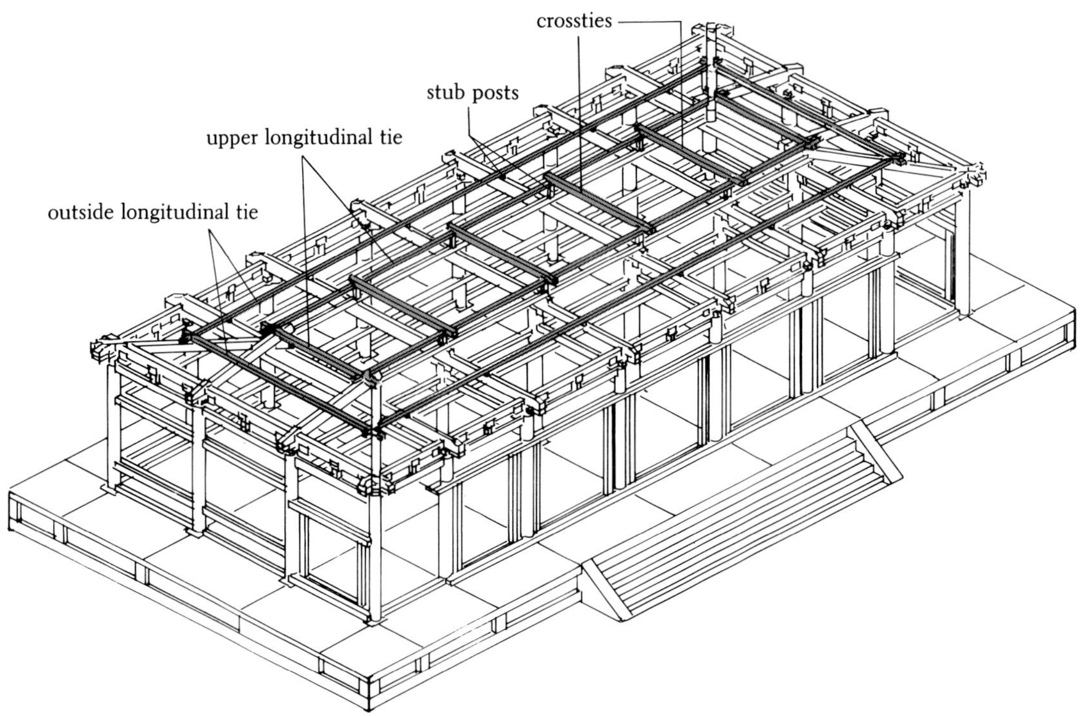

133. The Picture Hall after installment of stub posts, outside and upper longitudinal ties, and crossties.

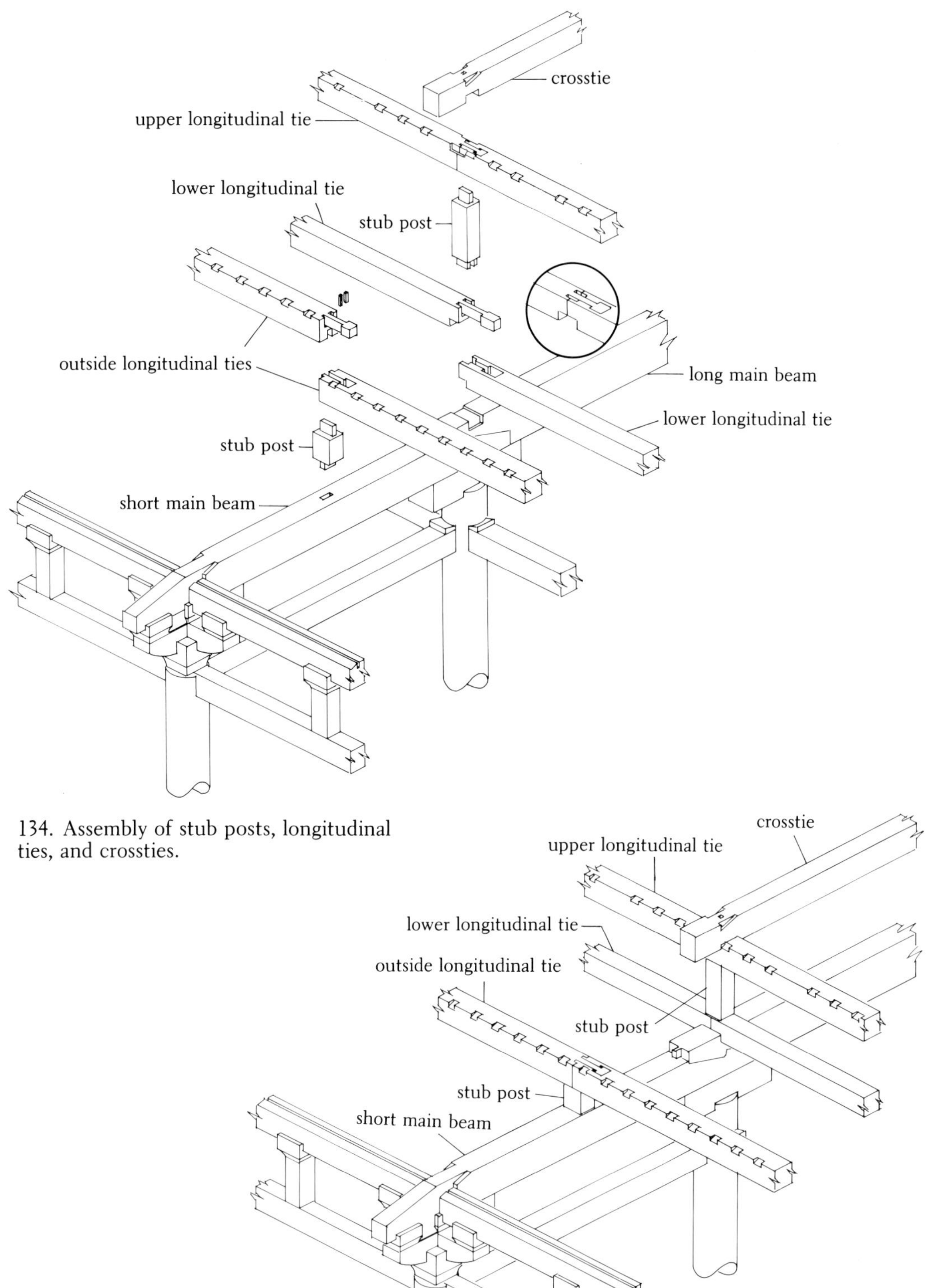

134. Assembly of stub posts, longitudinal ties, and crossties.

135. Completed assembly of stub posts, longitudinal ties, and crossties.

upper longitudinal tie

crosstie

lower longitudinal tie

long main beam

head tie

hidden beam support block

inner column

mid-wall tie

outside longitudinal tie

short main beam

136. Structure above typical inner column.

137. This stub post is located over a "gooseneck" splice joint in the longitudinal tie. Double tenons are used to straddle the "gooseneck" tenon below.

Hip Rafters

Hip rafters have two parts, lower (or "base") and upper (or "flying"). They cantilever outward beyond the wall line almost a third of their length and bear a significant proportion of the overhanging roof load (fig. 138). The lower hip rafter fits into slots at three points: where the outside and upper longitudinal ties overlap, on the corner main beam, and where the wall purlin and the corner bracket set come together (fig. 139, color pl. 7). It is shaped so as to lock these several layers of structure together while being supported by them.

The lower hip rafters require painstaking installation, as the ends of all four must lie exactly level to each other. Any discrepancies among the numerous supporting members must be resolved at this stage through careful trimming and adjustment, complicated by their great weight (figs. 141, 144). Once the lower hip rafter is in position, the upper portion can be simply installed, kept in place by a long groove and two rectangular pegs (figs. 145–46). The two will eventually be permanently jointed by large U-shaped wrought-iron clamping pins, but are left unconnected throughout the fitting of the rafters and rafter supports in the next section.

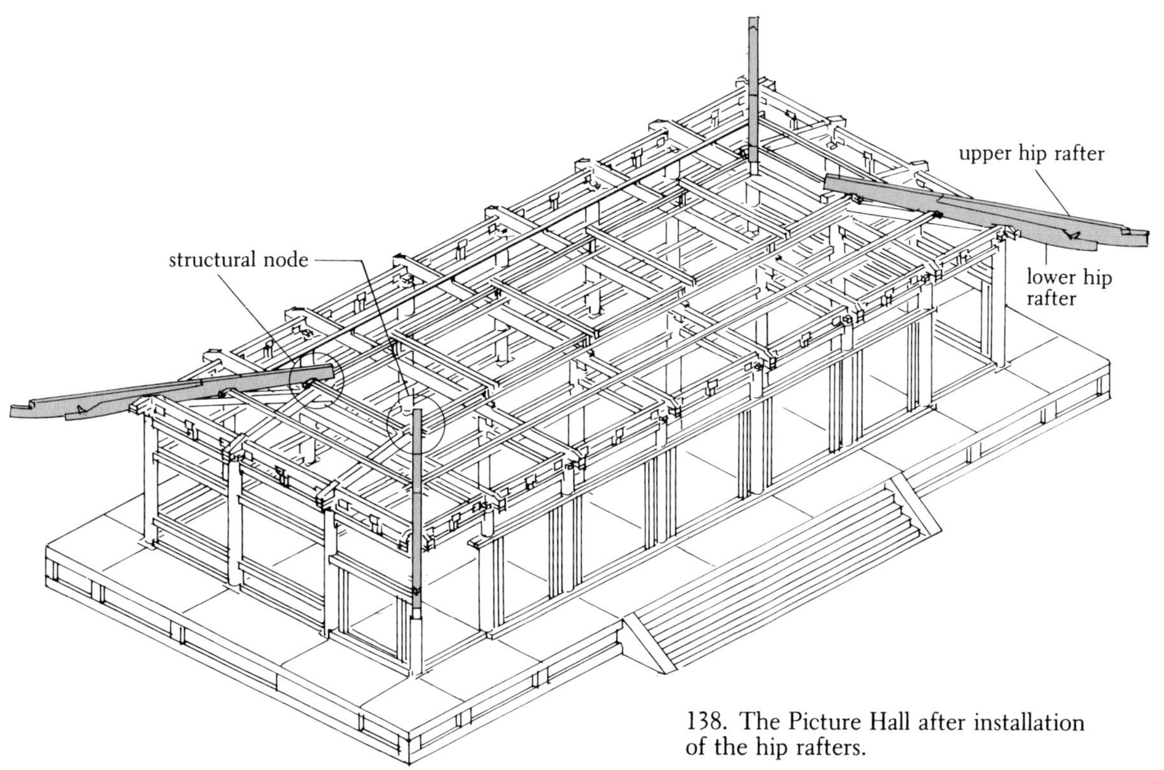

138. The Picture Hall after installation of the hip rafters.

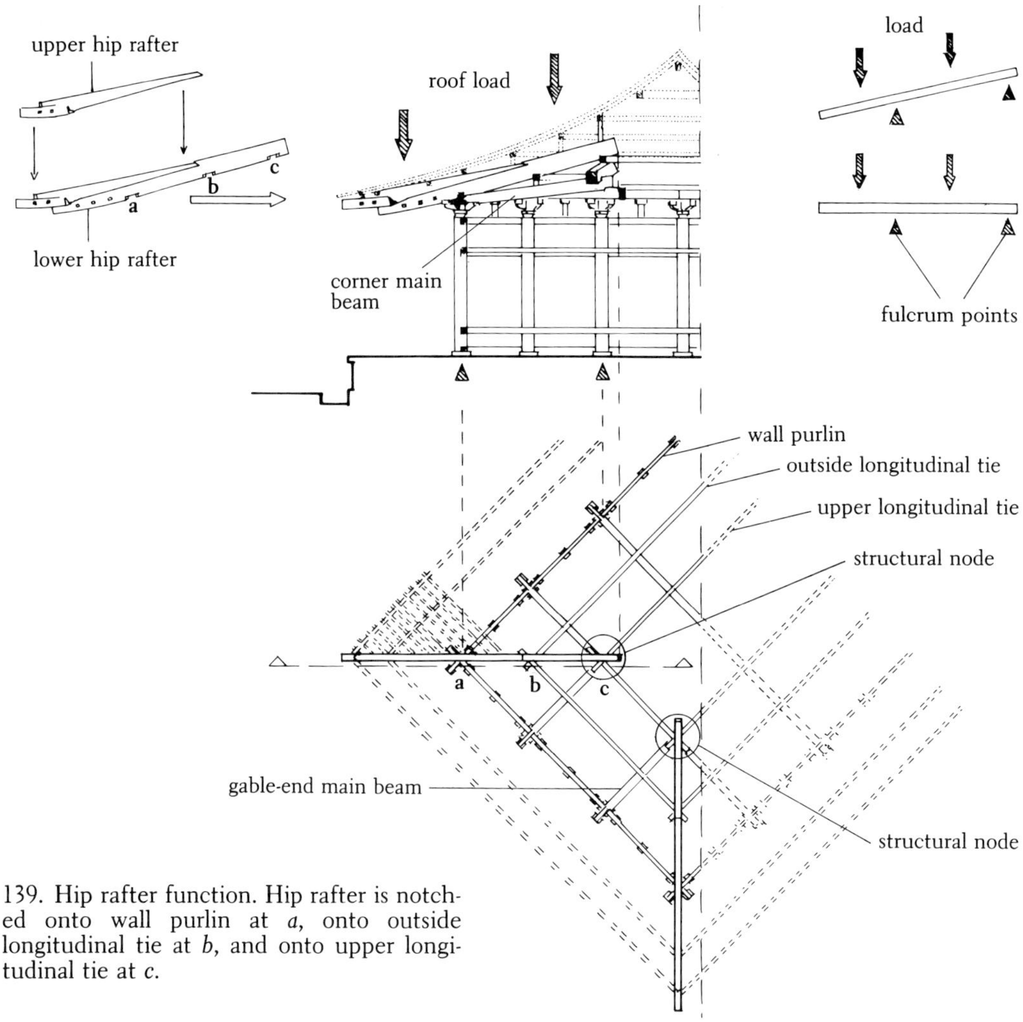

upper hip rafter

lower hip rafter

a b c

roof load

corner main beam

load

fulcrum points

wall purlin

outside longitudinal tie

upper longitudinal tie

structural node

a b c

gable-end main beam

structural node

139. Hip rafter function. Hip rafter is notched onto wall purlin at *a*, onto outside longitudinal tie at *b*, and onto upper longitudinal tie at *c*.

140. Lower hip rafter extends past corner bracket complex.

141. Lower hip rafter.

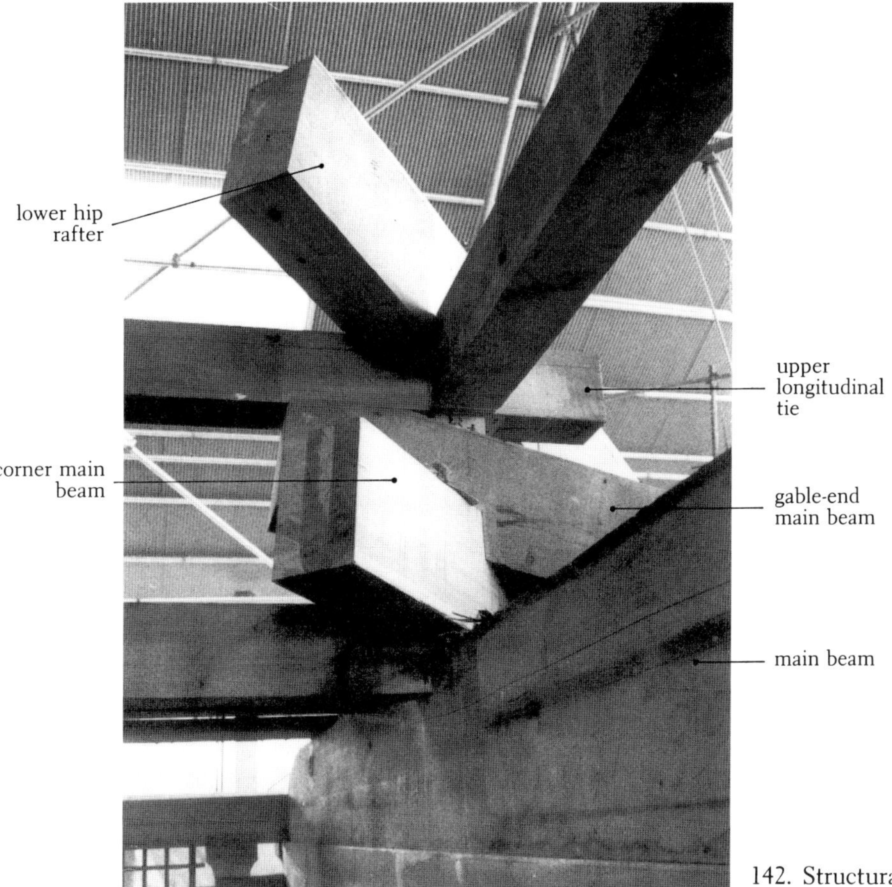

142. Structural node where beams and hip rafter meet.

143. The upper hip rafter as installed.

144. The fit of the lower hip rafter is repeatedly tested and trimmed if necessary.

145. The upper hip rafter will be mated to the lower hip rafter by means of square pegs and a long tongue-and-groove. See also figure 146.

146. Upper hip rafters upside down, showing long tongue which fits into corresponding groove in lower hip rafter. See also figure 145.

Lower Rafters and Upper Rafter Supports

The lower rafters lie across the outside and upper longitudinal ties and the wall purlin (fig. 148). Like the hip rafters, they cantilever out about two meters. The lower rafters nestle into toothlike slots on the longitudinal ties and into toothed boards set into a slot in the wall purlin. At their extreme ends, the lower rafters support a toothed beam which will in turn support the upper rafters (figs. 149–50). This beam, or upper rafter support, is made in sections; the rafters which support the joints are the first to be installed, their heights adjusted until the support beam is perfectly curved. (The upper rafter supports are in fact curved in both plan and elevation; that is, they twist upward and outward as they approach their intersection with the hip rafters; p. 66). Once the proper curve is established, the remaining lower rafters are installed and sandwiched into place at their upper ends by a hold-down beam. The rafters at the corner intersect with and are connected to the hip rafters, and so are shorter than the rest. Other rafters run alongside the main beams, and come in contact with the bracket sets. The rafters are secured in place with large wrought-iron spikes.

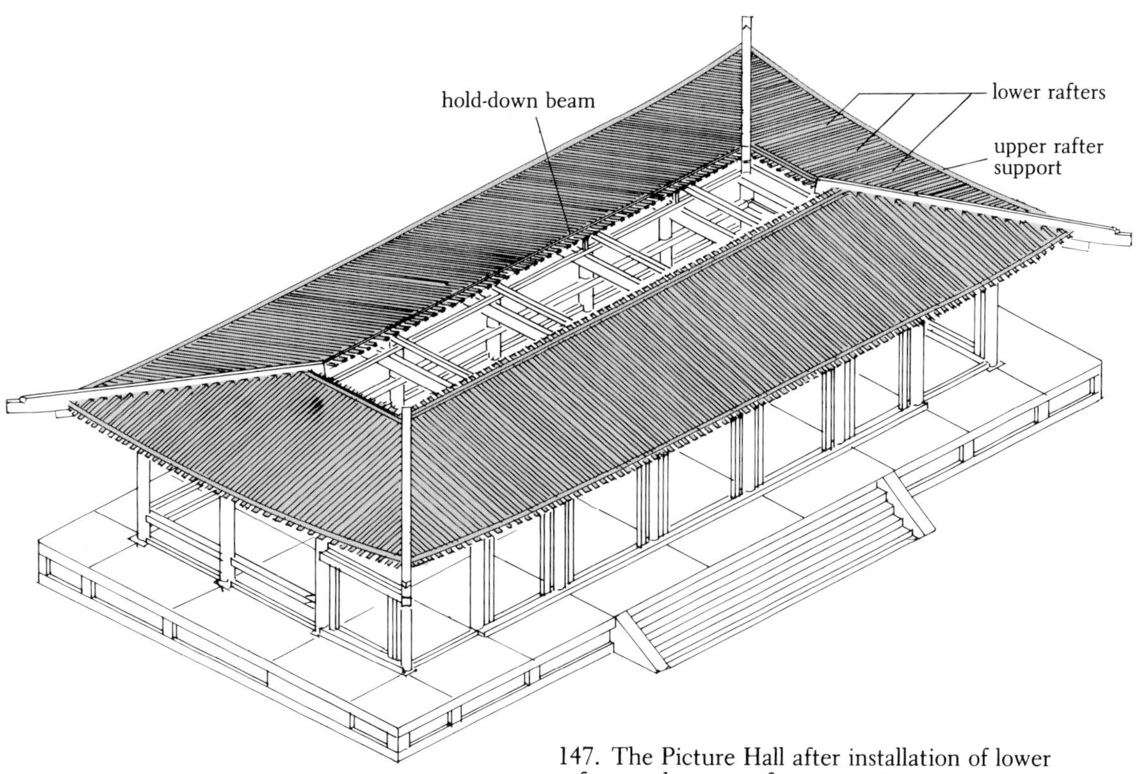

147. The Picture Hall after installation of lower rafters and upper rafter support.

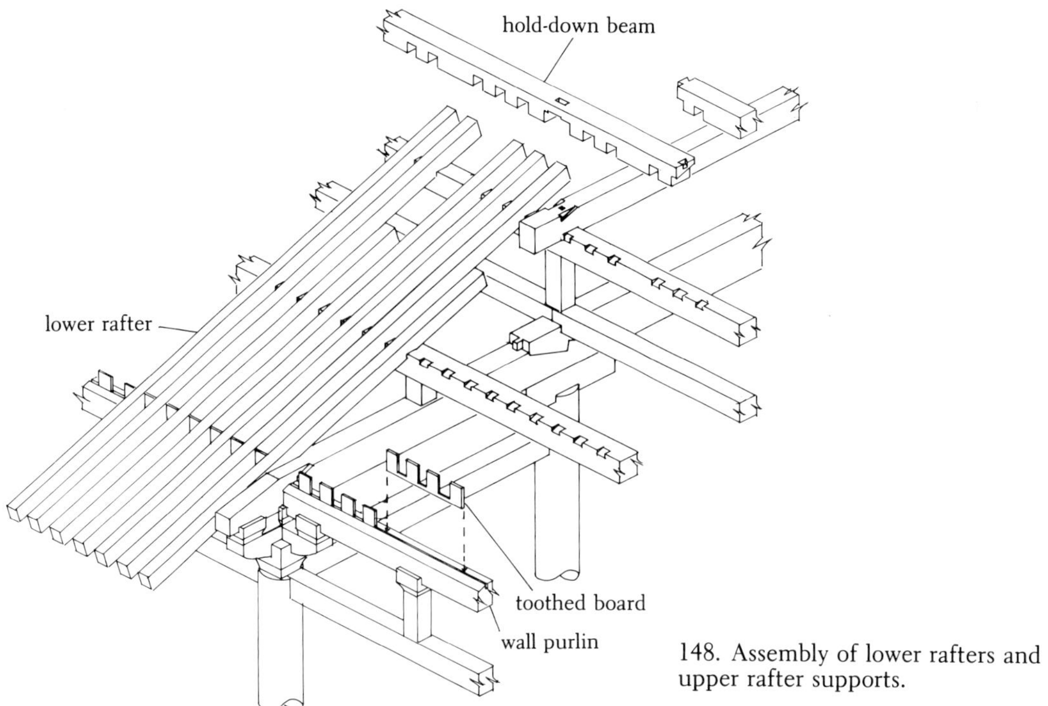

148. Assembly of lower rafters and upper rafter supports.

149. The upper rafter supports are themselves supported at the ends by the hip rafters. Their curve is very pronounced in both plan and section. See also figure 36.

150. The toothed upper rafter support is attached to the lower rafters by means of wrought-iron spikes.

Upper Rafters and Eaves Supports

Before the upper rafters can be installed (fig. 151), a layer of backing boards must be placed over the lower rafters (fig. 152). These are painted white on the underside and span the gaps between rafters, one board for each opening. They are positioned so as to allow expansion. The upper rafters are then inserted into the toothed slots in the upper rafter support, with their other tapered ends resting on the lower-rafter backing boards. The upper rafters support a curved beam similar to the upper-rafter support called the eaves support (fig. 153). The eaves support fills the gap between the upper rafters and the roofing tiles, and receives the lower ends of another layer of sheathing. A special joint is used to connect its segments. This beam is installed similarly to the upper rafter support, that is, with the jointed areas being supported and adjusted first. With the installation of the remaining rafters, the exposed underside portion of the roof is complete.

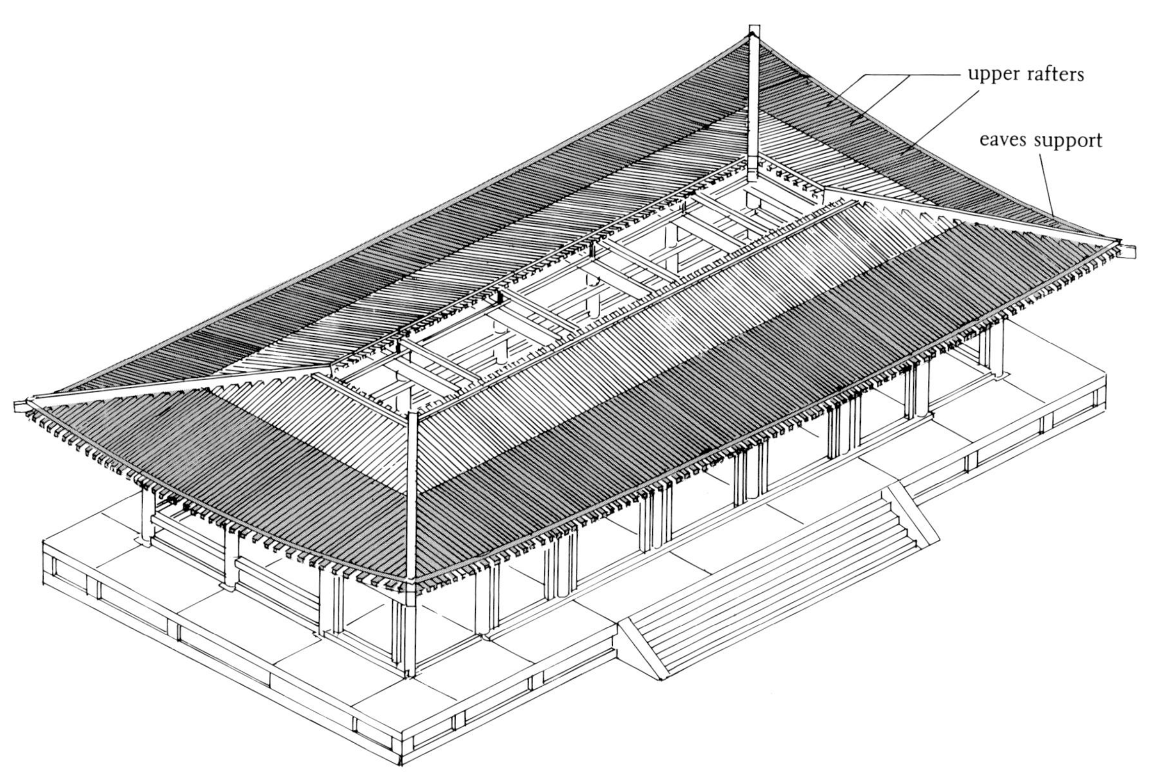

151. The Picture Hall after installation of upper rafters and eaves supports.

152. Assembly of upper rafters and eaves support. Inset shows dovetail splice joint on upper rafter support.

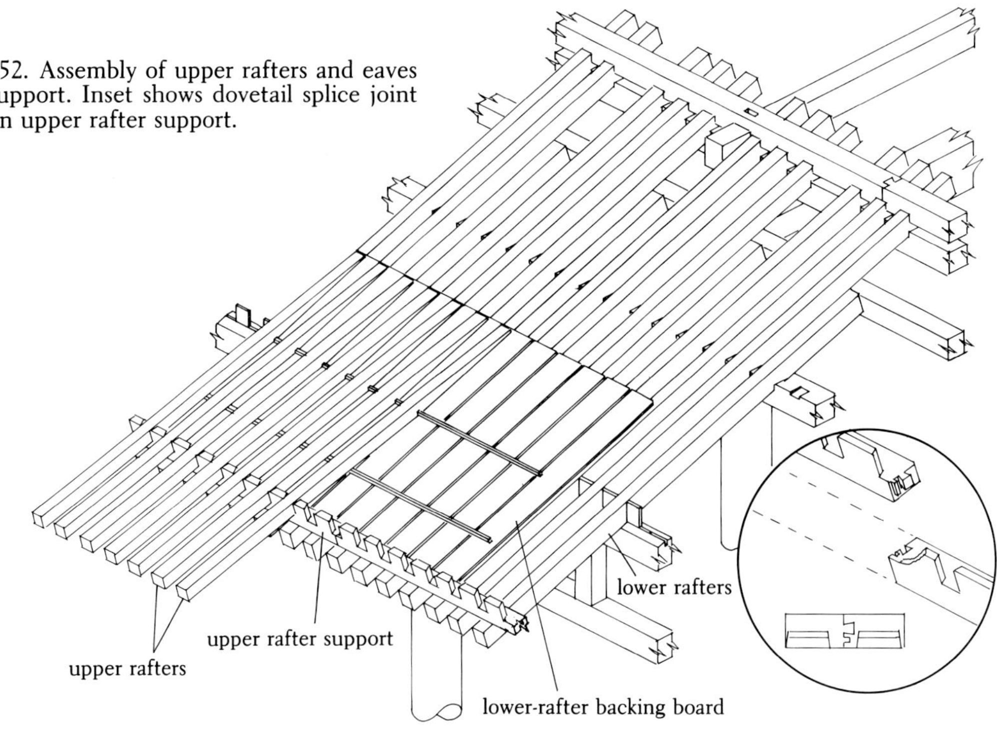

153. Assembly of upper-rafter backing board and eaves support. Inset shows splice joint of eaves support.

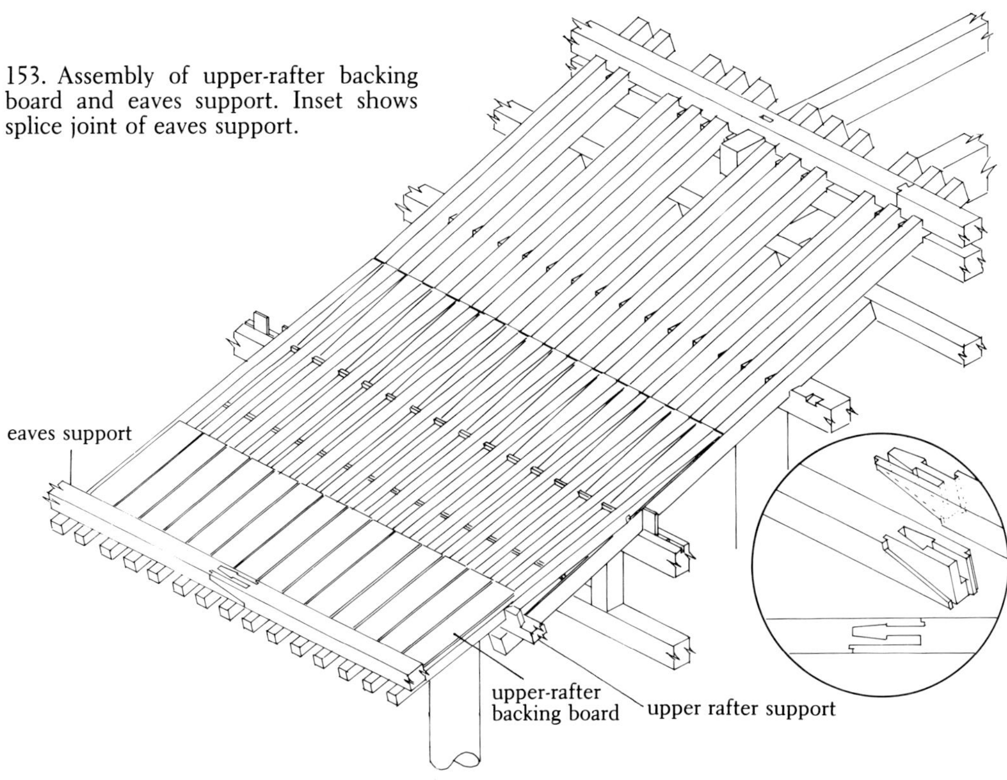

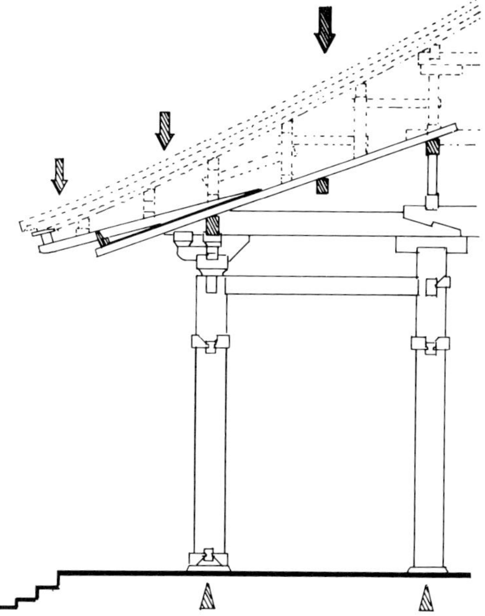

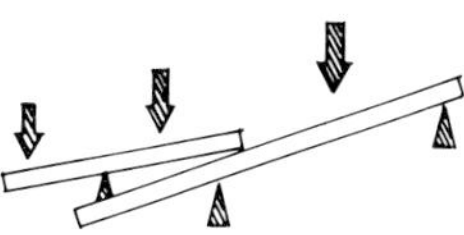

154. Cantilever action of rafters.

155. Upper rafters in relation to upper rafter support and eaves support.

156. The eaves support.

Hidden Roof Struts and Beams

One design feature that separates Japanese temples from their continental prototypes is the use of an illusion in the roof area. The rafters that are exposed at the eaves are not the "real" rafters in that they do not directly support the roof surface. The slope of the actual roof is steeper than indicated by the members exposed at the eaves; thus Japanese temples are said to have a "hidden" roof. In effect, a hollow space is formed between the gently sloping exposed rafters and the steeply sloping hidden rafters. This hollow space is filled with stub posts and roof purlins, and numerous small braces, or bridging (fig. 157). The actual rafters rest atop the purlins and are entirely hidden from view.

The stub posts are inserted into mortises in the rafter hold-down beams, of which there are six sets in all (fig. 158). The posts increase in height as they approach the roof ridge. Near the top of the roof, another set of crossties is installed, resting on blocks set atop the stub posts (fig. 159). All stub posts have slots for bridging. Two more sets of posts are inserted into the crossties, followed by two more roof purlins. This is followed by another set of crossties, and finally the blocks which support the hidden ridge beam itself. In addition, decorative purlins are installed which will be exposed at the gables (fig. 162, color pl. 8).

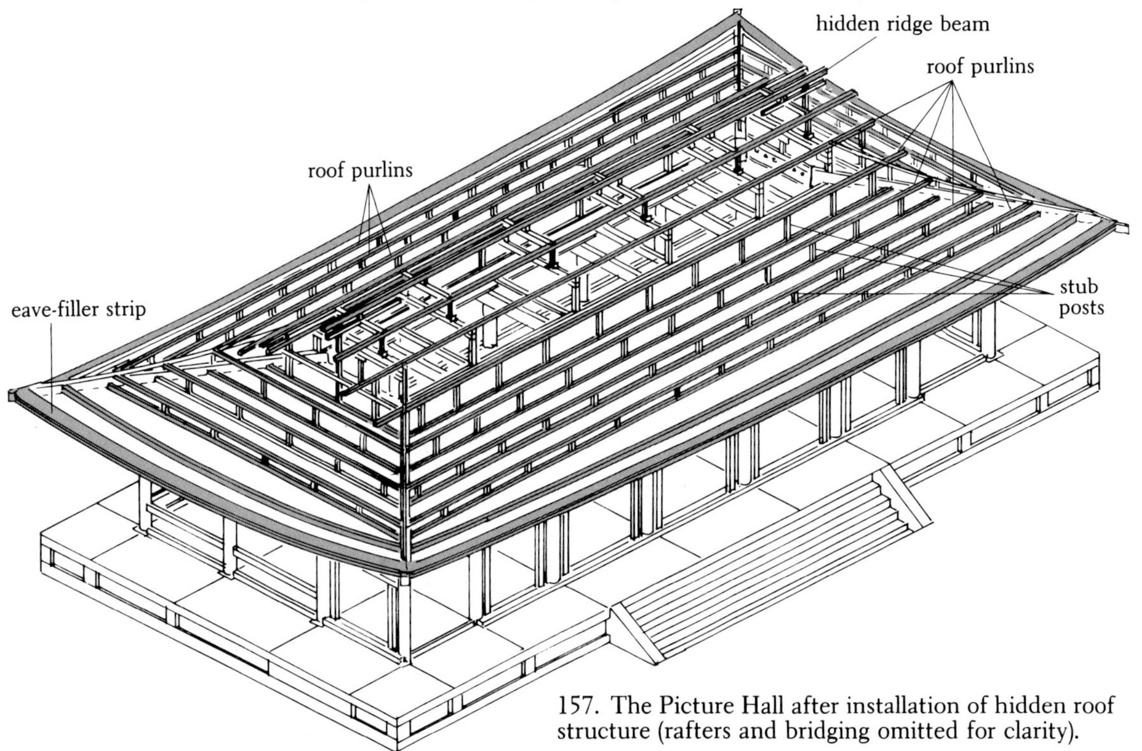

157. The Picture Hall after installation of hidden roof structure (rafters and bridging omitted for clarity).

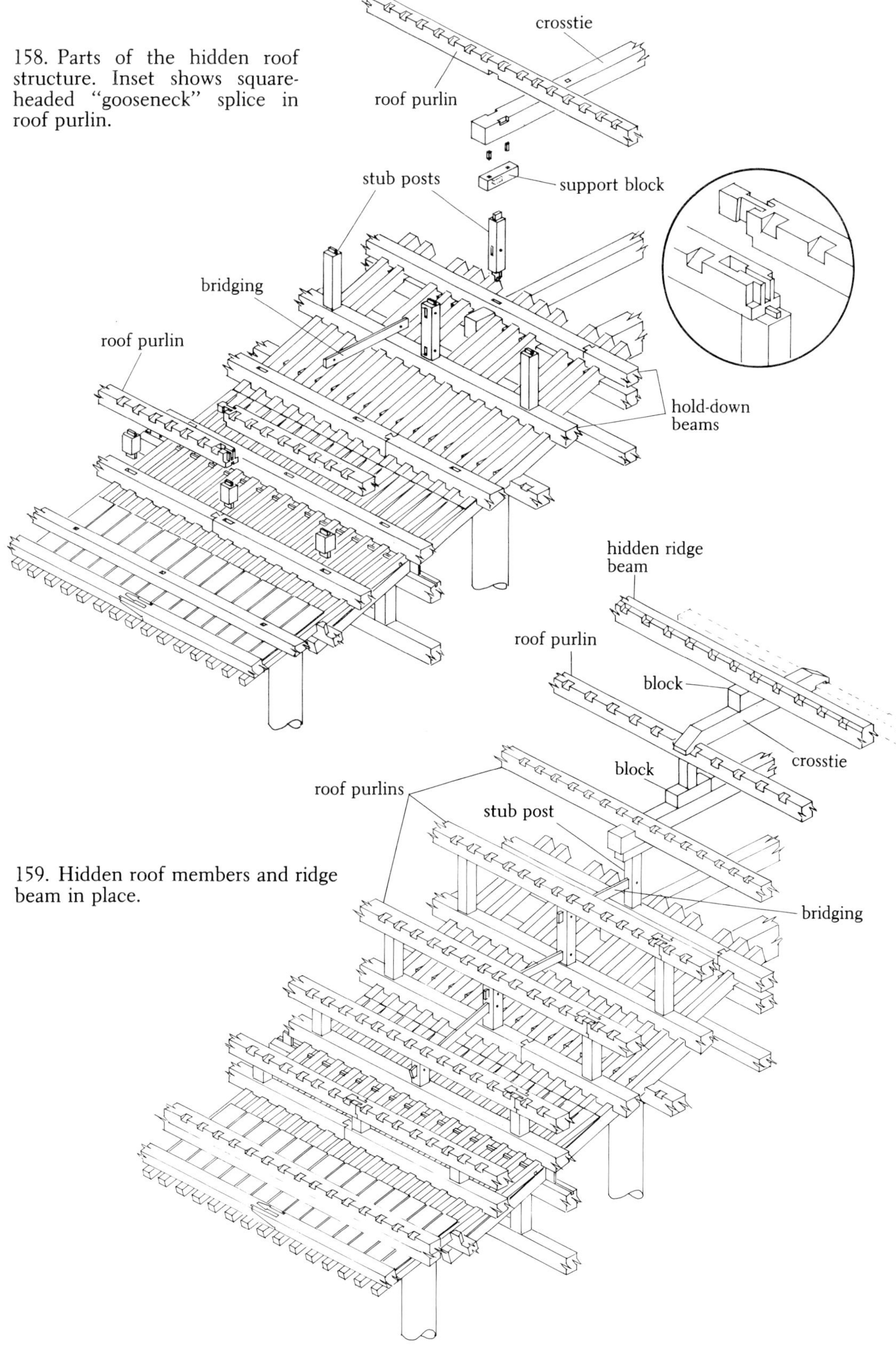

158. Parts of the hidden roof structure. Inset shows square-headed "gooseneck" splice in roof purlin.

159. Hidden roof members and ridge beam in place.

160. Uppermost roof purlins and stub posts installed.

161. The mid-crossties are carried on short blocks on top of the uppermost stub posts.

162. The hidden ridge beam is uppermost; below it are the short decorative beams that will be visible at the gable.

Finishing the Roof Structure

The hidden rafters are now installed above the roof purlins (figs. 163–66). They are noticeably smaller in section than the exposed rafters and must bend to accommodate the roof curve. Above the hidden rafters are laid the roofing boards. The gables, which will be exposed, are completed by the addition of fascia, a decorative set of crossbeams, struts, and diagonal braces (fig. 167), rafters, and a decorative pendant.

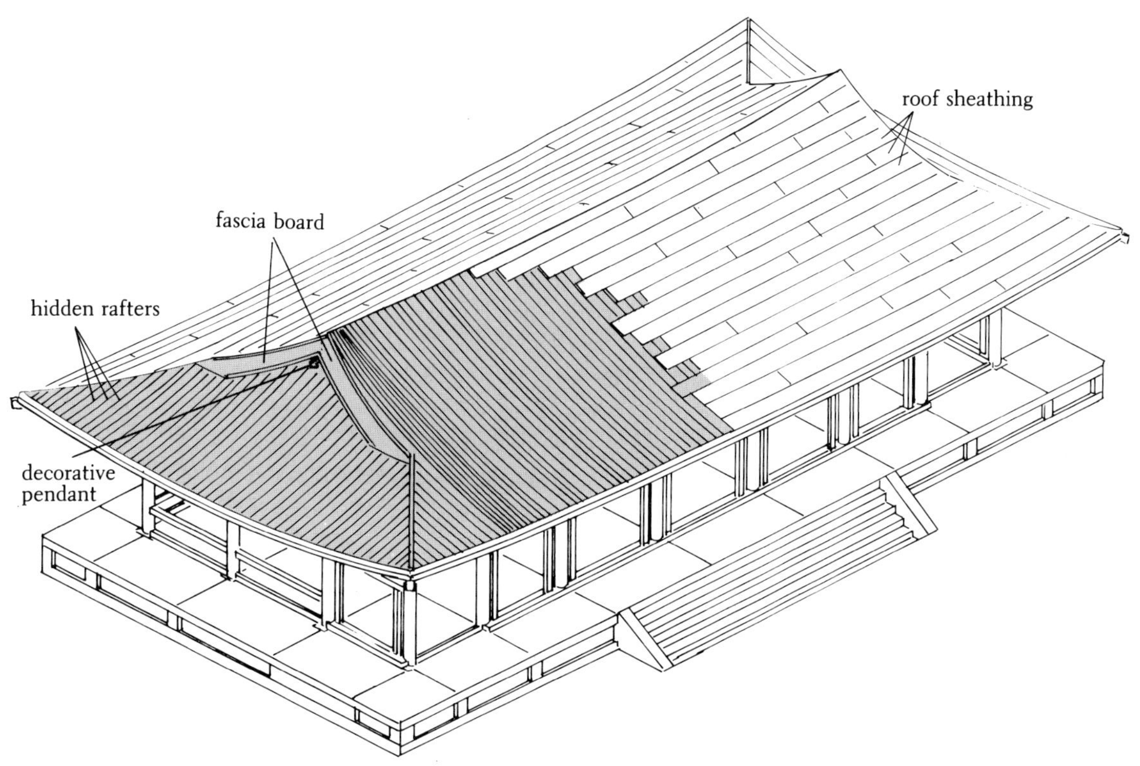

163. The Picture Hall after the installation of the hidden rafters and sheathing.

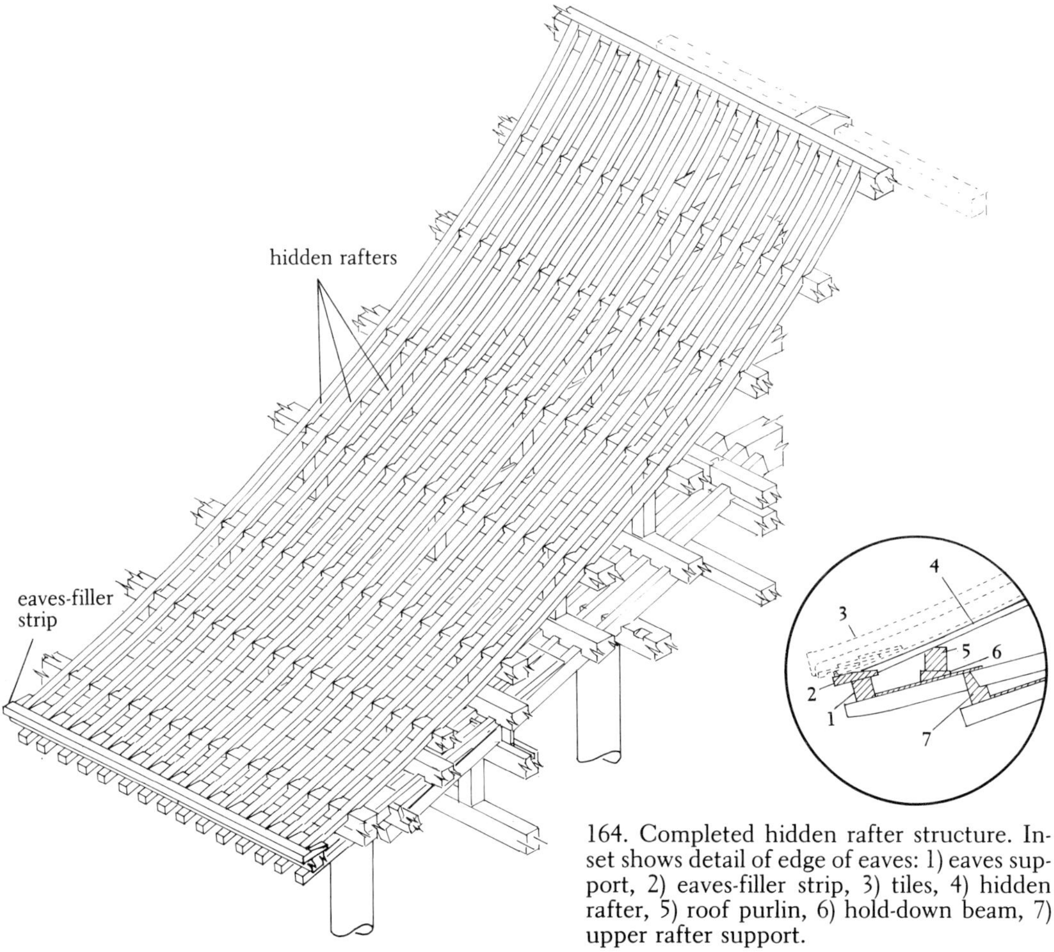

hidden rafters

eaves-filler strip

164. Completed hidden rafter structure. Inset shows detail of edge of eaves: 1) eaves support, 2) eaves-filler strip, 3) tiles, 4) hidden rafter, 5) roof purlin, 6) hold-down beam, 7) upper rafter support.

165. Since the curve of the hidden rafters will define the visible roof line, they are carefully placed with the use of a template.

166. Hidden rafters (visible on right) are covered with sheathing boards.

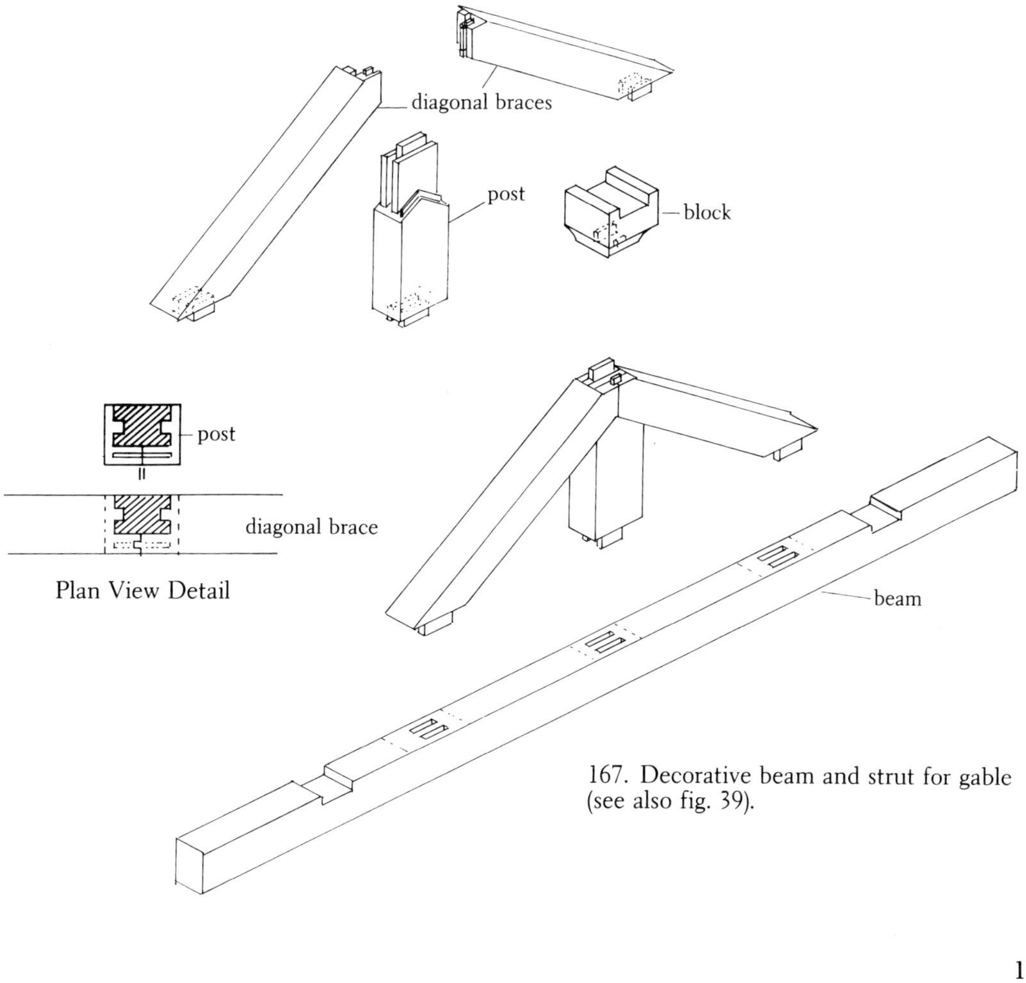

diagonal braces

post

block

post

diagonal brace

Plan View Detail

beam

167. Decorative beam and strut for gable (see also fig. 39).

Wattle and Daub Walls

Lath for the mud walls is installed all around (fig. 168). It consists of a lattice of thin wooden members which are inserted in shallow mortises in the columns and beams, and wrapped with rope (figs. 169–70). This forms a base to which the clay and straw plastering will adhere securely. The plaster is applied in several layers of slightly different formulation and allowed to dry thoroughly between coats. The first layers are applied by hand and the surface coats by trowel, then polished (figs. 171–72). The final wall is a hard, gleaming white (see color pl. 2; the gray areas will not be finished until the Picture Hall and surrounding corridor have been connected).

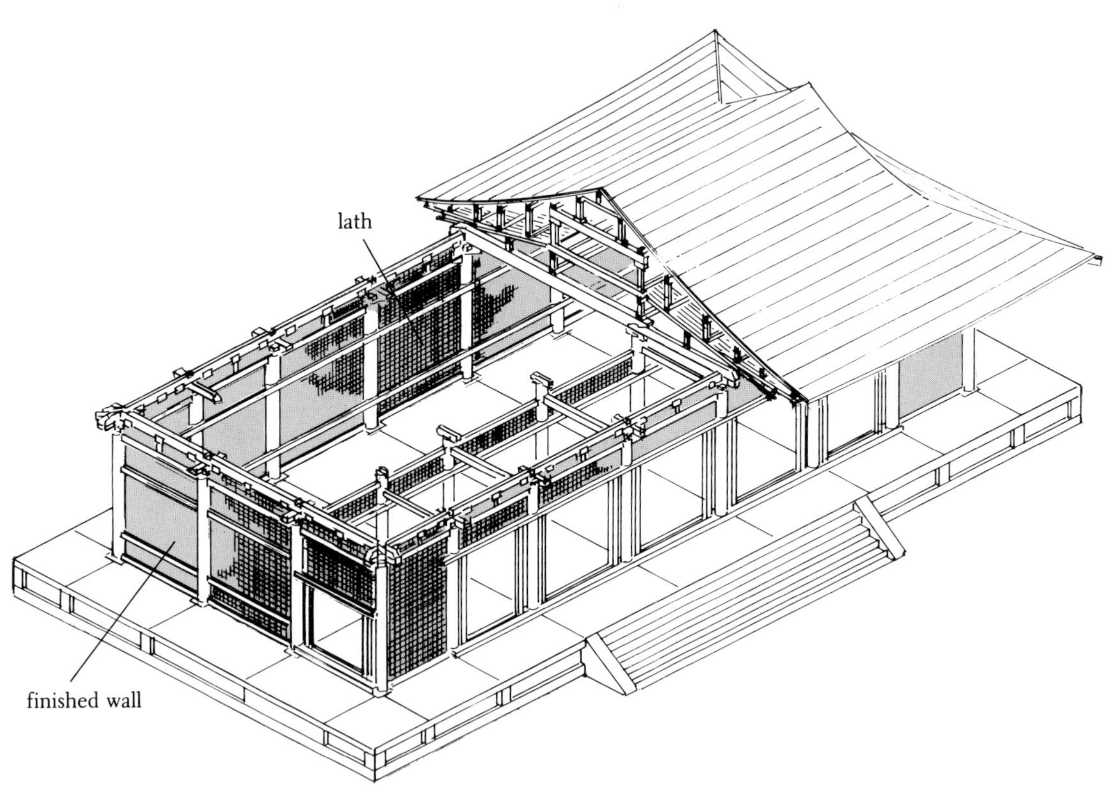

168. The Picture Hall in the process of wall installation.

169. Lath installed as a grid of thin wooden strips and wrapped with hemp rope.

170. Nearly completed latticework on the western wall.

171. Mud plaster as applied to the inner surface only.

172. Final coats are troweled on, creating a progressively smoother surface.

Ceiling

The ceiling is a jointed wooden lattice (fig. 173) surmounted by painted boards. It fits into the penetrating tie beams at the perimeter, and is supported over the center span by suspension rods attached to the roof structure. The grid over the outer zone is smaller than that over the inner zone and is noticeably lower. Hatchways are provided for access.

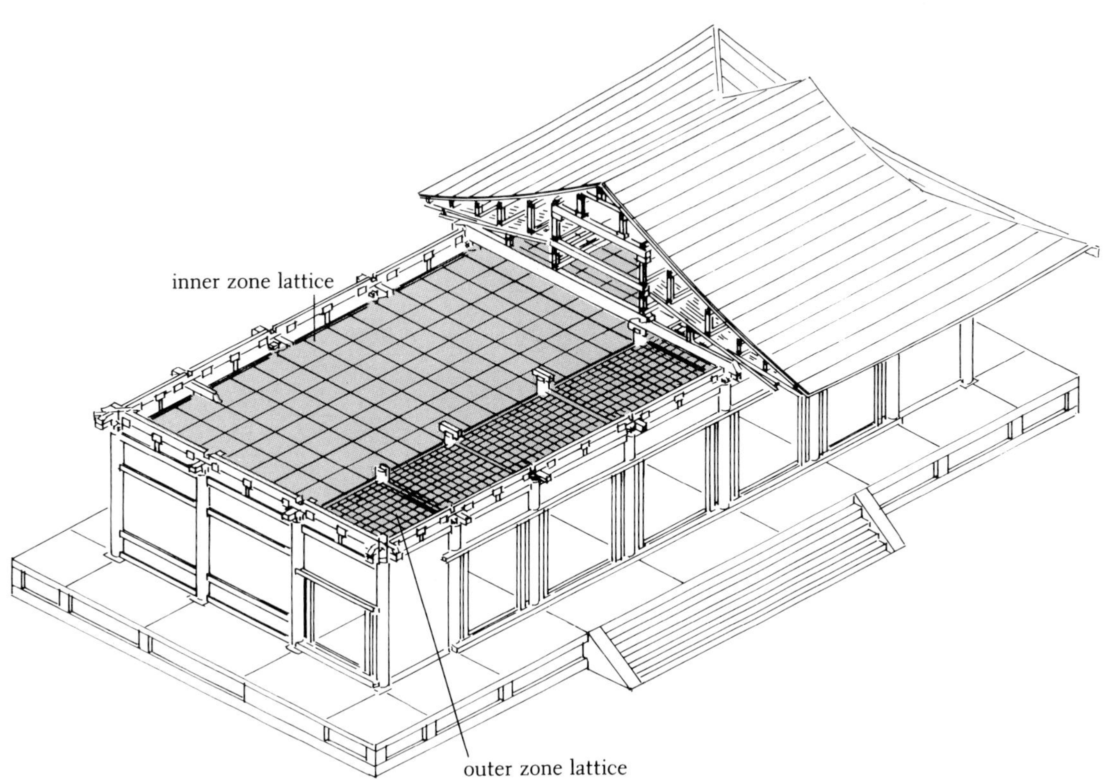

173. The Picture Hall after installation of ceiling lattice.

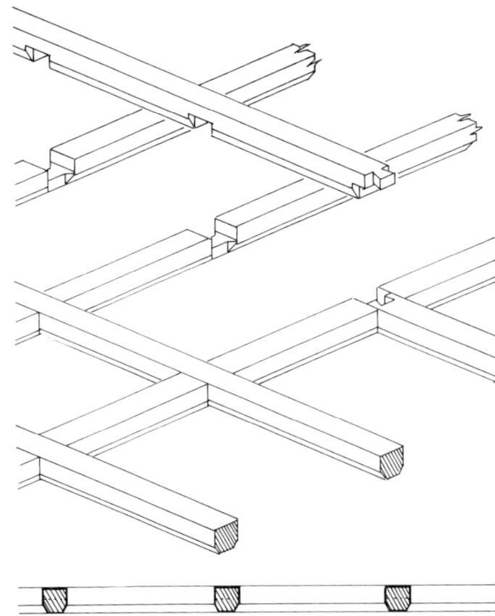

174. The ceiling is a grid of wooden members (approx. 10 cm square in section) which are jointed at every intersection.

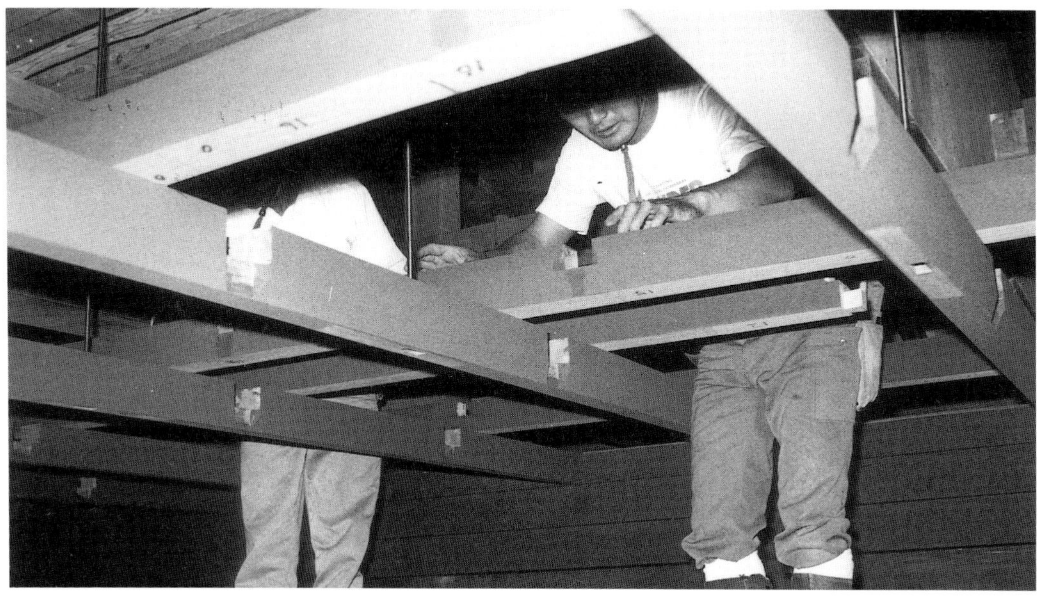

175. Positioning the ceiling members.

176. Over the outer zone, no suspension rods are required, because of the relatively short span.

Exterior Details

Doors are among the last elements to be added (color pl. 2; figs. 177–80). There are five doorways across the front, and one at each side (eventually to lead into the surrounding corridor); each doorway has a pair of doors. An individual door has five wooden parts: three thick panels (tongue-and-grooved) and two rails (upper and lower) which lock the panels together. The doors are decorated with thick gilt-bronze strap hardware.

The concrete foundation is completely faced and paved in granite (figs. 181–83). The facing is applied in a conventional manner, with adhesive mortar and metal straps embedded in the concrete. The paving is laid in a simple pattern, cut on the site to accommodate the column base stones. Mortar is applied dry, leveled, and then wet with a ladle. The stone is positioned with the help of small shims which ensure even spacing, and more water is applied and allowed to enter the cracks. The stone is tapped with a rubber mallet until perfectly level.

All of the rafters receive pierced-work bronze plates on their ends (color pl. 1; fig. 184). Bronze bells are hung from the hip rafters, and bronze bosses are applied to conceal spike heads on the outside horizontal brace beams.

The exterior walls are plastered white (color pl. 2), and the interior awaits the historical paintings which will be applied to the walls and the ceiling, as well as a warm-colored natural tile floor.

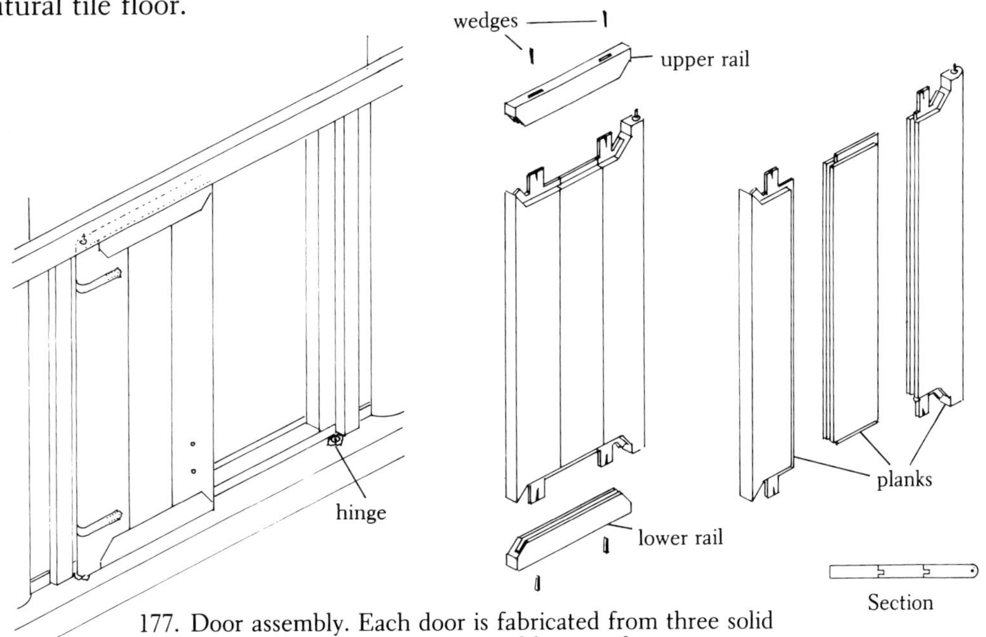

177. Door assembly. Each door is fabricated from three solid planks, locked together by upper and lower rails.

178. Finished doors are almost airtight.

179. Smooth pivot hinges of gilt bronze.

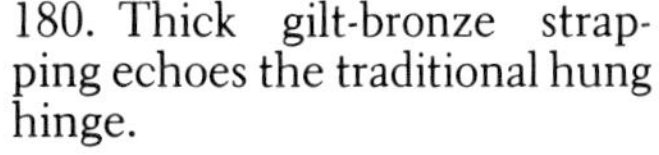

180. Thick gilt-bronze strapping echoes the traditional hung hinge.

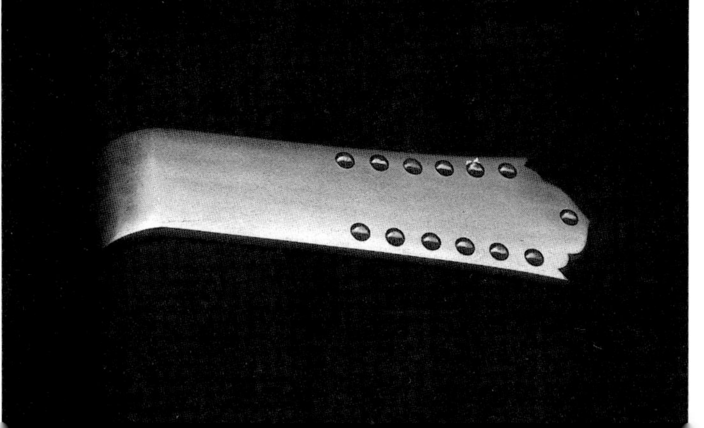

181. Granite facing and paving stones are among the last details to be added.

182. Paving stones are set in a thick bed of mortar.

183. Detail of paving stones and column base stone.

184. Applying gilt-bronze pierced-work rafter end caps.

Roof Tiles

Traditionally, Japanese architecture uses several types of roof-tiling systems. The buildings at Yakushiji use a silver-gray tile manufactured nearby, with a decorative pattern taken from the original eighth-century pagoda. This roof requires tiles of several shapes.

The primary tiles are slightly dished rectangles and are placed in rows placed side to side, with upper tiles overlapping lower.(figs. 188–91).

Their joint lines are covered with semi-cylindrical tiles, which overlap the tiles on either side (figs. 192–94). All of the roof ridges are covered by stacked rectangular tiles, crown side up, which are in turn capped by the semi-cylindrical tiles. These groups are terminated by large decorative "ogre" tiles and bird-perch tiles (fig. 198).

Before the modern era, the space between the curve of the roof sheathing and the tiles was often filled with mud, which made the curve of the roof smooth and provided a secure bed for the tiles. At Yakushiji, considerable weight is saved by laying the tiles on wooden battens, fastened in crucial places by nails and copper ties. The resultant seal is equally as watertight as the traditional method. This completes the roof.

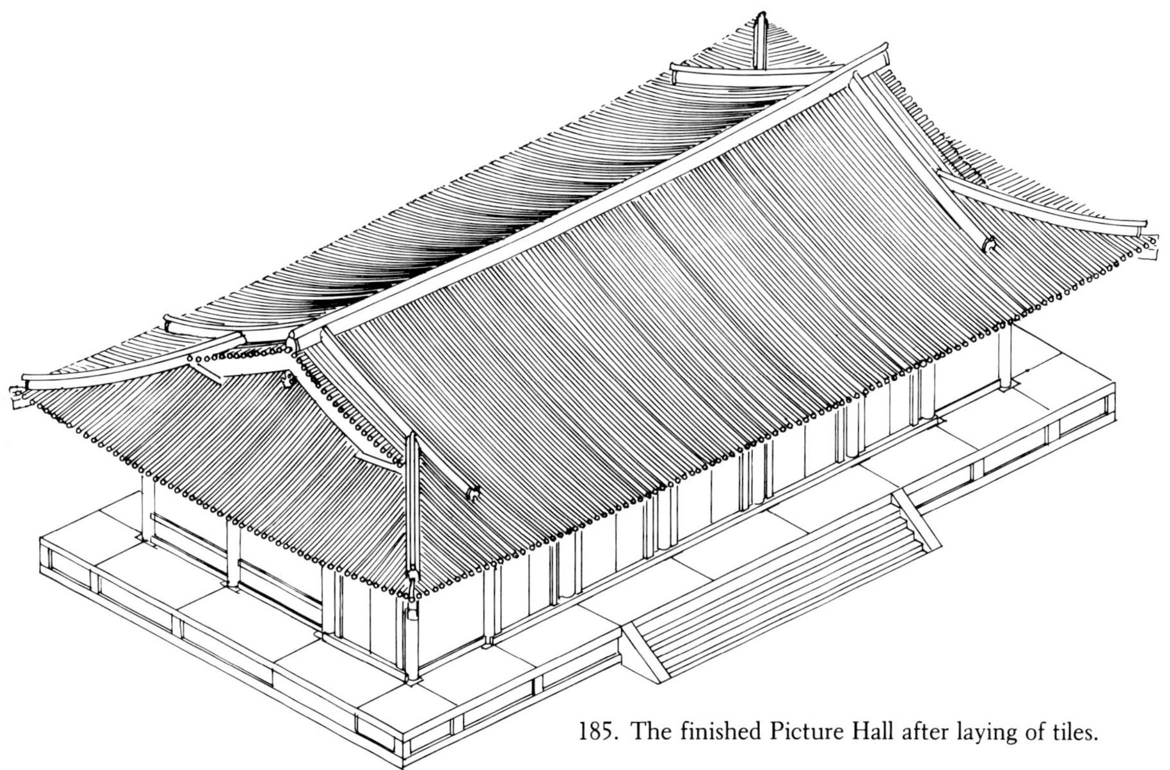

185. The finished Picture Hall after laying of tiles.

186. A layer of cedar bark is applied over the roof sheathing boards, and then a set of battens are laid horizontally.

187. Another set of battens are laid vertically, and a scalloped tile cant strip is added.

188. The primary eave tiles, which are decorated at the end, are the first to be laid, fastened with wire ties to the vertical battens.

189. The remaining dished primary tiles are applied systematically, row by row.

190. There is a small gap between adjacent rows of primary tiles.

191. The roof after all the primary tiles have been laid.

192. The gap between primary tiles is covered by half-cylindrical tiles. Those at the eaves are decorated.

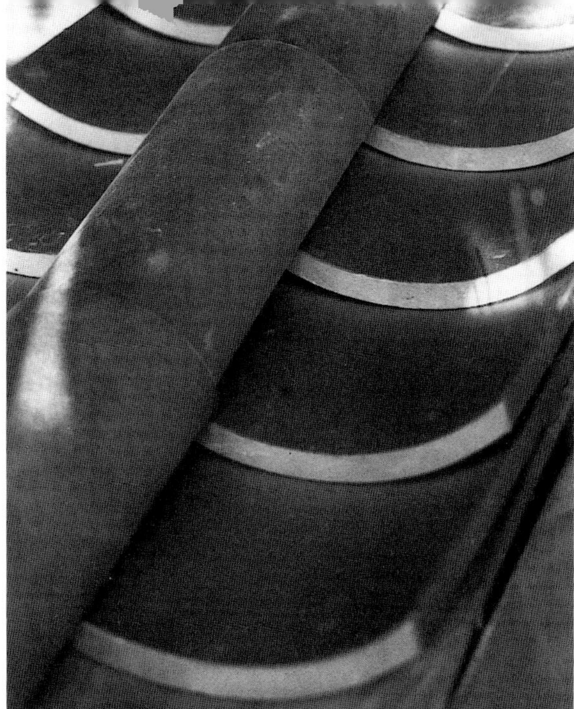

193. The half-cylindrical tiles are laid so that upper overlaps lower.

194. Half-cylindrical tiles in place.

195. Ridges are formed by stacked, slightly convex tiles set in mortar.

196. Ridge tiles are held in place with small wooden spacers until the mortar has set.

197. The ridges are capped with half-cylindrical tiles. Angled intersections are covered with specially shaped tiles.

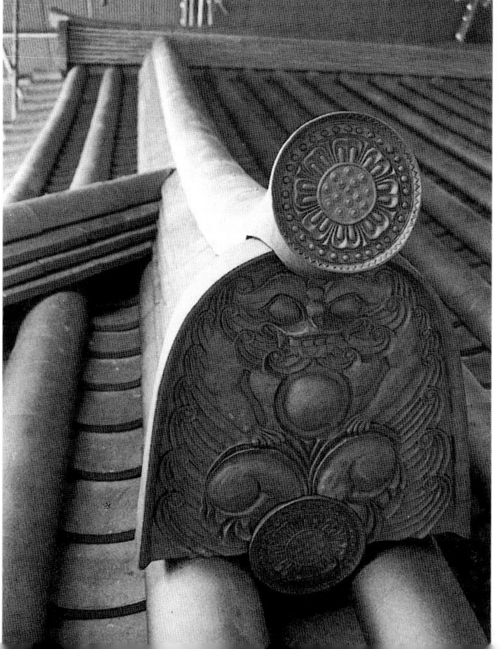

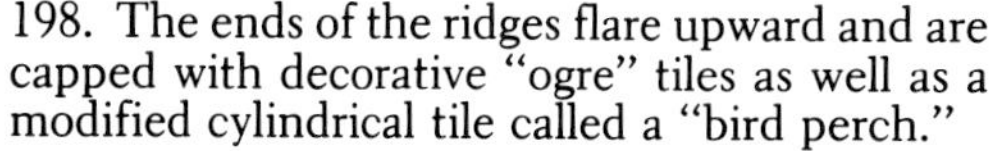

198. The ends of the ridges flare upward and are capped with decorative "ogre" tiles as well as a modified cylindrical tile called a "bird perch."

THE RIDGE-BEAM CEREMONY

One highlight of the construction process is the raising of the uppermost beam. It is a cause for celebration. True, the job is not entirely finished yet. A thousand details need attending to over the following months, but the structure is finally a building. The roof has holes, but it is a roof, and the spirits can reside beneath it. This is the meaning of the ridge-beam–raising ceremony (see figs. 199–205). The old spirits that dwell on the property—be they rice-field spirits or forest spirits—are kindly asked to vacate the premises, and are given numerous inducements to do so in the form of food and gifts. The spirits of heaven then lower the ridge beam, suitably adorned, into place. These new deities do not take up residence as yet; that occurs at the formal enshrinement ceremony after everything is really finished. But the building is now a temple.

Reflecting the fact that in Japan there is a peaceful coexistence between indigenous Shintō deities and their imported Buddhist counterparts from China, the ridge-beam raising combines practices from both faiths. The carpenters, accordingly, dress in the vestments of Shintō priests for the day, led by the master in appeasing the Shintō spirits of the earth and wind with songs and gifts. Buddhist priests, meanwhile, read the sutras and make offerings of incense. Then, at the climactic moment, each priest holds one end of a long, colored banner, symbolically dangling the ridge beam from heaven, while carpenters above tap the beam into place with ritual tools made specifically for the occasion. All then retire to an adjoining hall for a celebratory feast, including copious quantities of sake served for once by the masters to the lowly.

Nowhere, perhaps, can the Japanese carpenter's enthusiasm be felt as keenly as at these celebrations, which rarely occur more than once in a year. Here the mood is jubilant and thankful: everyone concerned congratulating themselves and each other on exacting work well done, and thankful to all—gods, men, and fate—who have provided the opportunity to do work worthy of human beings, worthy of trained craftsmen, and to take part in the sheer joy of building.

199. A carpenter arranging flowers on an altar in preparation for the ridge-beam raising ceremony.

200. Offerings include sake, vegetables, rice cakes, and bitter oranges, as well as exquisitely decorated ceremonial tools.

201. Special talismans are made and temporarily installed, such as this *karimata* arrow pointing to heaven.

202. The ceremony is led by the abbot of the temple, Kōin Takada.

203. For this day, the carpenters are dressed as Shintō priests. Their job is to communicate with the spirits which reside in the wood.

204. Nishioka, as head carpenter, leads the ceremony on the roof while the Buddhist monks wait below. The apprentices wield ceremonial mallets to "tap" the ridge beam into place.

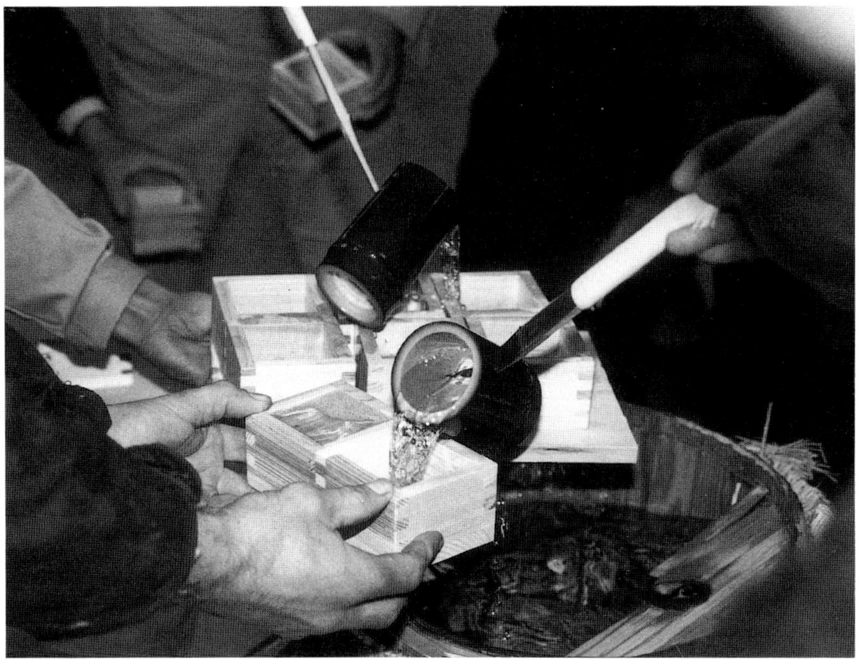

205. The ceremony concluded, casks of sake are broken open, square wooden cups filled, and the meal begun—punctuated by songs, speeches, and the occasionally sound of a thoroughly satisfied carpenter slipping off his inebriated chair and hitting the earthen floor with a thud.

FURTHER READING

Engel, Heinrich. *Measure and Construction of the Japanese House.* Tokyo and Rutland, Vermont: C. E. Tuttle, 1985.
Extracted from the author's larger work, *The Japanese House: A Tradition for Contemporary Architecture,* this book has copious illustrations dealing with house planning and construction.

Nakahara, Yasuo. *Japanese Joinery: A Handbook for Joiners and Carpenters.* Translated by Koichi Paul Nii. Point Roberts, Washington: Cloudburst Press/Hartley and Marks, 1983.
A fairly detailed discussion of Japanese carpentry, including joinery, but dealing exclusively with house construction. Still, a valuable book, and the best of its kind in English.

Nakashima, George. *The Soul of a Tree: A Woodworker's Reflections.* Tokyo and New York: Kodansha International, 1981.
A poetic and evocative essay on trees and the woodworker's ethic by the renowned Japanese-American cabinetmaker.

Nishi, Kazuo, and Kazuo Hozumi. *What Is Japanese Architecture?* Translated by H. Mack Horton. Tokyo and New York: Kodansha International, 1985.
A good overall introduction to Japanese architecture, including temples, the history of Nara, and other background pertinent to temple architecture. Contains a good bibliography.

Odate, Toshio. *Japanese Woodworking Tools: Their Tradition, Spirit, and Use.* Newtown, Conn.: Taunton Press, 1984.
One of the few books in English dealing with Japanese tools, a "how-to" book. Good descriptions of the various saws, planes, chisels, etc.

Ooka, Minoru. *Temples of Nara and Their Art.* Translated by Dennis Lishka. (Heibonsha Survey of Japanese Art, vol. 7). Tokyo and New York: Weatherhill/Heibonsha, 1973.
A detailed study of Nara Buddhist architecture, with special attention paid to the evolution of layouts and some information about proportions.

Ōta Hirotarō, ed. *Japanese Architecture and Gardens.* Tokyo: Kokusai Bunka Shinkokai, 1966.
A fine, detailed introduction to Japanese architecture, with particular attention paid to temple construction.

Parent, Mary Neighbour. *The Roof in Japanese Buddhist Architecture.* Tokyo and New York: Weatherhill/Kajima Shuppankai, 1983.
The best and most thorough analysis of Japanese temple roof structures, from the Asuka through the Muromachi periods. Extensive supplementary information.

Suzuki, Kakichi. *Early Buddhist Architecture in Japan.* Translated by Mary N. Parent and Nancy S. Steinhardt. (Japan Arts Library, vol. 9). Tokyo and New York: Kodansha International and Shibundo, 1980.
A detailed and thorough introduction to Buddhist architecture from the Asuka to the Heian periods. Much technical information and bibliography.

INDEX

The terminology of temple architecture presents many difficulties, one of which is that the Japanese terms often have no standard translation. In the text I have attempted to give English equivalents wherever possible, while noting Japanese counterparts in the index. My apologies to readers who would prefer that the Japanese be more readily accessible.

Kyoto
Kyushu
Shikoku
Osaka
Nara